Leveled Texts
for Mathematics

Measurement

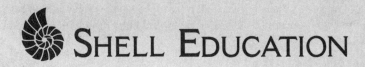

Author

Christi Sorell

SHELL EDUCATION

Consultant

Barbara Talley, M.S.
Texas A&M University

Publishing Credits

Dona Herweck Rice, *Editor-in-Chief*; Lee Aucoin, *Creative Director*; Don Tran, *Print Production Manager*;
Sara Johnson, M.S.Ed., *Senior Editor*; Hillary Wolfe, *Editor*; Stephanie Paris, *Editor*;
Evelyn Garcia, *Associate Education Editor*; Neri Garcia, *Cover Designer*; Juan Chavolla, *Production Artist*;
Stephanie Reid, *Photo Editor*; Corinne Burton, M.S.Ed., *Publisher*

All images from Shutterstock.com

Shell Education

5301 Oceanus Drive

Huntington Beach, CA 92649

http://www.shelleducation.com

ISBN 978-1-4258-0754-2

©2011 Shell Educational Publishing, Inc.

Table of Contents

What Is Differentiation?

Over the past few years, classrooms have evolved into diverse pools of learners. Gifted students, English language learners, special needs students, high achievers, underachievers, and average students all come together to learn from one teacher. The teacher is expected to meet their diverse needs in one classroom. It brings back memories of the one-room schoolhouse during early American history. Not too long ago, lessons were designed to be one size fits all. It was thought that students in the same grade level learned in similar ways. Today, we know that viewpoint to be faulty. Students have differing learning styles, come from different cultures, experience a variety of emotions, and have varied interests. For each subject, they also differ in academic readiness. At times, the challenges teachers face can be overwhelming, as they struggle to figure out how to create learning environments that address the differences they find in their students.

What is differentiation? Carol Ann Tomlinson at the University of Virginia says, "Differentiation is simply a teacher attending to the learning needs of a particular student or small group of students, rather than teaching a class as though all individuals in it were basically alike" (2000). Differentiation can be carried out by any teacher who keeps the learners at the forefront of his or her instruction. The effective teacher asks, "What am I going to do to shape instruction to meet the needs of all my learners?" One method or methodology will not reach all students.

Differentiation encompasses what is taught, how it is taught, and the products students create to show what they have learned. When differentiating curriculum, teachers become the organizers of learning opportunities within the classroom environment. These categories are often referred to as content, process, and product.

- **Content:** Differentiating the content means to put more depth into the curriculum through organizing the curriculum concepts and structure of knowledge.

- **Process:** Differentiating the process means using varied instructional techniques and materials to enhance students' learning.

- **Product:** When products are differentiated, cognitive development and the students' abilities to express themselves improve, as they are given different product options.

Teachers should differentiate content, process, and product according to students' characteristics, including students' readiness, learning styles, and interests.

- **Readiness:** If a learning experience aligns closely with students' previous skills and understanding of a topic, they will learn better.

- **Learning styles:** Teachers should create assignments that allow students to complete work according to their personal preferences and styles.

- **Interests:** If a topic sparks excitement in the learners, then students will become involved in learning and better remember what is taught.

4

How to Differentiate Using This Product

According to the Common Core State Standards (2010), all students need to learn to read and discuss concepts across the content areas in order to be prepared for college and beyond. The leveled texts in this series help teachers differentiate mathematics content for their students to allow all students access to the concepts being explored. Each book has 15 topics, and each topic has a text written at four different reading levels. (See page 17 for more information.) While these texts are written at a variety of reading levels, all the levels remain strong in presenting the mathematics content and vocabulary. Teachers can focus on the same content standard or objective for the whole class, but individual students can access the content at their instructional reading levels rather than at their frustration levels.

Determining your students' instructional reading levels is the first step in the process. It is important to assess their reading abilities often so they do not get tracked into one level. Below are suggested ways to determine students' reading levels.

- **Running records:** While your class is doing independent work, pull your below-grade-level students aside, one at a time. Have them read aloud the lowest level of a text (the star level) individually as you record any errors they make on your own copy of the text. If students read accurately and fluently and comprehend the material, move them up to the next level and repeat the process. Following the reading, ask comprehension questions to assess their understanding of the material. Use your judgment to determine whether students seem frustrated as they read. As a general guideline, students reading below 90% accuracy are likely to feel frustrated as they read. There are also a variety of published reading assessment tools that can be used to assess students' reading levels with the running record format.

- **Refer to other resources:** Other ways to determine instructional reading levels include checking your students' Individualized Education Plans (IEP), asking the school's resource teachers, or reviewing test scores. All of these resources should be able to give you the additional information you need to determine the reading level to begin with for your students.

Teachers can also use the texts in this series to scaffold the content for their students. At the beginning of the year, students at the lowest reading levels may need focused teacher guidance. As the year progresses, teachers can begin giving students multiple levels of the same text to allow them to work independently to improve their comprehension. This means each student would have a copy of the text at his or her independent reading level and instructional reading level. As students read the instructional-level texts, they can use the lower texts to better understand the difficult vocabulary. By scaffolding the content in this way, teachers can support students as they move up through the reading levels. This will encourage students to work with texts that are closer to the grade level at which they will be tested.

General Information About the Student Populations

Below-Grade-Level Students

By Dennis Benjamin

Gone are the days of a separate special education curriculum. Federal government regulations require that special needs students have access to the general education curriculum. For the vast majority of special needs students today, their Individualized Education Plans (IEPs) contain current and targeted performance levels but few short-term content objectives. In other words, the special needs students are required to learn the same content as their on-grade-level peers.

Be well aware of the accommodations and modifications written in students' IEPs. Use them in your teaching and assessment so they become routine. If you hold high expectations of success for all of your students, their efforts and performances will rise as well. Remember the root word of *disability* is *ability*. Go to the root needs of the learner and apply good teaching. The results will astound and please both of you.

English Language Learners

By Marcela von Vacano

Many school districts have chosen the inclusion model to integrate English language learners into mainstream classrooms. This model has its benefits as well as its drawbacks. One benefit is that English language learners may be able to learn from their peers by hearing and using English more frequently. One drawback is that these second-language learners cannot understand academic language and concepts without special instruction. They need sheltered instruction to take the first steps toward mastering English. In an inclusion classroom, the teacher may not have the time or necessary training to provide specialized instruction for these learners.

Acquiring a second language is a lengthy process that integrates listening, speaking, reading, and writing. Students who are newcomers to the English language are not able to process information until they have mastered a certain number of structures and vocabulary words. Students may learn social language in one or two years. However, academic language takes up to eight years for most students.

Teaching academic language requires good planning and effective implementation. Pacing, or the rate at which information is presented, is another important component in this process. English language learners need to hear the same word in context several times, and they need to practice structures to internalize the words. Reviewing and summarizing what was taught are absolutely necessary for English language learners.

General Information About the Student Populations

On-Grade-Level Students

By Wendy Conklin

On-grade-level students often get overlooked when planning curriculum. More emphasis is usually placed on those who struggle and, at times, on those who excel. Teachers spend time teaching basic skills and even go below grade level to ensure that all students are up to speed. While this is a noble endeavor and is necessary at times, in the midst of it all, the on-grade-level students can get lost in the shuffle. We must not forget that differentiated strategies are good for the on-grade-level students, too. Providing activities that are too challenging can frustrate these students; on the other hand, assignments that are too easy can be boring and a waste of their time. The key to reaching this population successfully is to find just the right level of activities and questions while keeping a keen eye on their diverse learning styles.

Above-Grade-Level Students

By Wendy Conklin

In recent years, many state and school district budgets have cut funding that has in the past provided resources for their gifted and talented programs. The push and focus of schools nationwide is proficiency. It is important that students have the basic skills to read fluently, solve problems, and grasp mathematical concepts. As a result, funding has been redistributed in hopes of improving test scores on state and national standardized tests. In many cases, the attention has focused only on improving low test scores to the detriment of the gifted students who need to be challenged.

Differentiating the products you require from your students is a very effective and fairly easy way to meet the needs of gifted students. Actually, this simple change to your assignments will benefit all levels of students in your classroom. While some students are strong verbally, others express themselves better through nonlinguistic representation. After reading the texts in this book, students can express their comprehension through different means, such as drawings, plays, songs, skits, or videos. It is important to identify and address different learning styles. By giving more open-ended assignments, you allow for more creativity and diversity in your classroom. These differentiated products can easily be aligned with content standards. To assess these standards, use differentiated rubrics.

 #50754—Leveled Texts for Mathematics: Measurement

Strategies for Using the Leveled Texts

Below-Grade-Level Students

By Dennis Benjamin

Vocabulary Scavenger Hunt

A valuable prereading strategy is a Vocabulary Scavenger Hunt. Students preview the text and highlight unknown words. Students then write the words on specially divided pages. The pages are divided into quarters with the following headings: *Definition*, *Sentence*, *Examples*, and *Nonexamples*. A section called *Picture* can be placed over the middle of the chart to give students a visual reminder of the word and its definition.

Example Vocabulary Scavenger Hunt

estimate

Definition	Sentence
to determine roughly the size or quantity	We need to estimate how much paint we will need to buy.
Examples	**Nonexamples**
estimating a distance; estimating an amount	measuring to an exact number; weighing and recording an exact weight

This encounter with new vocabulary enables students to use it properly. The definition identifies the word's meaning in student-friendly language. The sentence should be written so that the word is used in context. This helps the student make connections with background knowledge. Illustrating the sentence gives a visual clue. Examples help students prepare for factual questions from the teacher or on standardized assessments. Nonexamples help students prepare for ***not*** and ***except for*** test questions such as "All of these are polygons *except for…*" and "Which of these terms in this expression is not a constant?" Any information the student was unable to record before reading can be added after reading the text.

Strategies for Using the Leveled Texts *(cont.)*

Below-Grade-Level Students *(cont.)*

Graphic Organizers to Find Similarities and Differences

Setting a purpose for reading content focuses the learner. One purpose for reading can be to identify similarities and differences. This is a skill that must be directly taught, modeled, and applied. The authors of *Classroom Instruction That Works* state that identifying similarities and differences "might be considered the core of all learning" (Marzano, Pickering, and Pollock 2001). Higher-level tasks include comparing and classifying information and using metaphors and analogies. One way to scaffold these skills is through the use of graphic organizers, which help students focus on the essential information and organize their thoughts.

Example Classifying Graphic Organizer

Equation	Constants	Variables	Number of Terms
$7 + 12 = 19$	7, 12, 19	none	3
$3x = 12$	12	x	2
$a + b$	none	a, b	2
$a^2 + b^2 + c^2$	none	a, b, c	3

The Riddles Graphic Organizer allows students to compare and contrast two-dimensional shapes using riddles. Students first complete a chart you've designed. Then, using that chart, they can write summary sentences. They do this by using the riddle clues and reading across the chart. Students can also read down the chart and write summary sentences. With the chart below, students could write the following sentences: A circle is not a polygon. The interior angles of a triangle always add up to 180°.

Example Riddles Graphic Organizer

What Am I?	Circle	Square	Triangle	Rectangle
I come in different configurations.			x	x
I am a polygon.		x	x	x
I am a closed shape.	x	x	x	x
My interior angles always add up to 180°.			x	
I have at least three vertices.		x	x	x

9

Strategies for Using the Leveled Texts *(cont.)*

Below-Grade-Level Students *(cont.)*

Framed Outline

This is an underused technique that bears great results. Many below-grade-level students have problems with reading comprehension. They need a framework to help them attack the text and gain confidence in comprehending the material. Once students gain confidence and learn how to locate factual information, the teacher can fade out this technique.

There are two steps to successfully using this technique. First, the teacher writes cloze sentences. Second, the students complete the cloze activity and write summary sentences.

Example Framed Outline

A _____ graph is used to show how two variables may be related to each other. A graph should have a _____. The axes should be labeled and a proper scale should be shown.

Summary Sentences

A good graph should correctly show the data. It should have a title; the axes should be labeled correctly; it should show the proper scale.

Modeling Written Responses

A frequent criticism heard by special educators is that below-grade-level students write poor responses to content-area questions. This problem can be remedied if resource and classroom teachers model what good answers look like. While this may seem like common sense, few teachers take the time to do this. They just assume all children know how to respond in writing.

This is a technique you may want to use before asking your students to respond to the You Try It questions associated with the leveled texts in this series. First, read the question aloud. Then, write the question on the board or an overhead and think aloud about how you would go about answering the question. Next, solve the problem showing all the steps. Introduce the other problems and repeat the procedure. Have students explain how they solved the problems in writing so that they make the connection that quality written responses are part of your expectations.

Strategies for Using the Leveled Texts (cont.)

English Language Learners

By Marcela von Vacano

Effective teaching for English language learners requires effective planning. In order to achieve success, teachers need to understand and use a conceptual framework to help them plan lessons and units. There are six major components to any framework. Each is described in detail below.

1. Select and Define Concepts and Language Objectives—Before having students read one of the texts in this book, the teacher must first choose a mathematical concept and language objective (reading, writing, listening, or speaking) appropriate for the grade level. Then, the next step is to clearly define the concept to be taught. This requires knowledge of the subject matter, alignment with local and state objectives, and careful formulation of a statement that defines the concept. This concept represents the overarching idea. The mathematical concept should be written on a piece of paper and posted in a visible place in the classroom.

By the definition of the concept, post a set of key language objectives. Based on the content and language objectives, select essential vocabulary from the text. The number of new words selected should be based on students' English language levels. Post these words on a word wall that may be arranged alphabetically or by themes.

2. Build Background Knowledge—Some English language learners may have a lot of knowledge in their native language, while others may have little or no knowledge. The teacher will want to build the background knowledge of the students using different strategies such as the following:

Visuals: Use posters, photographs, postcards, newspapers, magazines, drawings, and video clips of the topic you are presenting.

Realia: Bring real-life objects to the classroom. If you are teaching about measurement, bring in items such as thermometers, scales, time pieces, and rulers.

Vocabulary and Word Wall: Introduce key vocabulary in context. Create families of words. Have students draw pictures that illustrate the words and write sentences about the words. Also, be sure you have posted the words on a word wall in your classroom.

Desk Dictionaries: Have students create their own desk dictionaries using index cards. On one side, they should draw a picture of the word. On the opposite side, they should write the word in their own language and in English.

English Language Learners *(cont.)*

3. Teach Concepts and Language Objectives—The teacher must present content and language objectives clearly. He or she must engage students using a hook and must pace the delivery of instruction, taking into consideration students' English language levels. The concept or concepts to be taught must be stated clearly. Use the first languages of the students whenever possible or assign other students who speak the same languages to mentor and to work cooperatively with the English language learners.

Lev Semenovich Vygotsky, a Russian psychologist, wrote about the Zone of Proximal Development (ZPD). This theory states that good instruction must fill the gap that exists between the present knowledge of a child and the child's potential (1978). Scaffolding instruction is an important component when planning and teaching lessons. English language learners cannot jump stages of language and content development. You must determine where the students are in the learning process and teach to the next level using several small steps to get to the desired outcome. With the leveled texts in this series and periodic assessment of students' language levels, teachers can support students as they climb the academic ladder.

4. Practice Concepts and Language Objectives—English language learners need to practice what they learn with engaging activities. Most people retain knowledge best after applying what they learn to their own lives. This is definitely true for English language learners. Students can apply content and language knowledge by creating projects, stories, skits, poems, or artifacts that show what they learned. Some activities should be geared to the right side of the brain. For students who are left-brain dominant, activities such as defining words and concepts, using graphic organizers, and explaining procedures should be developed. The following teaching strategies are effective in helping students practice both language and content:

> **Simulations**: Students learn by doing. For example, when teaching about data analysis, have students do a survey about their classmates' favorite sports. First, students make a list of questions and collect the necessary data. Then, they tally the responses and determine the best way to represent the data. Lastly, students create a graph that shows their results and display it in the classroom.

> **Literature response**: Read a text from this book. Have students choose two concepts described or introduced in the text. Ask students to create a conversation two people might have to debate which concept is useful. Or, have students write journal entries about real-life ways they use these mathematical concepts.

Strategies for Using the Leveled Texts (cont.)

English Language Learners (cont.)

4. Practice Concepts and Language Objectives (cont.)

Have a short debate: Make a controversial statement such as "It isn't necessary to learn addition." After reading a text in this book, have students think about the question and take a position. As students present their ideas, one student can act as a moderator.

Interview: Students may interview a member of the family or a neighbor in order to obtain information regarding a topic from the texts in this book. For example: What are some ways you use geometry in your work?

5. Evaluation and Alternative Assessments—We know that evaluation is used to inform instruction. Students must have the opportunity to show their understanding of concepts in different ways and not only through standard assessments. Use both formative and summative assessments to ensure that you are effectively meeting your content and language objectives. Formative assessment is used to plan effective lessons for a particular group of students. Summative assessment is used to find out how much the students have learned. Other authentic assessments that show day-to-day progress are: text retelling, teacher rating scales, student self-evaluations, cloze testing, holistic scoring of writing samples, performance assessments, and portfolios. Periodically assessing student learning will help you ensure that students continue to receive the correct levels of texts.

6. Home-School Connection—The home-school connection is an important component in the learning process for English language learners. Parents are the first teachers, and they establish expectations for their children. These expectations help shape the behavior of their children. By asking parents to be active participants in the education of their children, students get a double dose of support and encouragement. As a result, families become partners in the education of their children and chances for success in your classroom increase.

You can send home copies of the texts in this series for parents to read with their children. You can even send multiple levels to meet the needs of your second-language parents as well as your students. In this way, you are sharing your mathematics content standards with your whole second-language community.

13

Strategies for Using the Leveled Texts *(cont.)*

Above-Grade-Level Students

By Wendy Conklin

Open-Ended Questions and Activities

Teachers need to be aware of activities that provide a ceiling that is too low for gifted students. When given activities like this, gifted students become bored. We know these students can do more, but how much more? Offering open-ended questions and activities will give high-ability students the opportunities to perform at or above their ability levels. For example, ask students to evaluate mathematical topics described in the texts, with questions such as "Do you think students should be allowed to use calculators in math?" or "What do you think you would need to build a two-story dog house?" These questions require students to form opinions, think deeply about the issues, and form several different responses in their minds. To questions like these, there really is no single correct answer.

The generic, open-ended question stems listed below can be adapted to any topic. There is one You Try It question for each topic in this book. Use questions or statements like the ones shown here to develop further discussion for the leveled texts.

- In what ways did…
- How might you have done this differently…
- What if…
- What are some possible explanations for…
- How does this affect…
- Explain several reasons why…
- What problems does this create…
- Describe the ways…
- What is the best…
- What is the worst…
- What is the likelihood…
- Predict the outcome…
- Form a hypothesis…
- What are three ways to classify…
- Support your reason…
- Make a plan for…
- Propose a solution…
- What is an alternative to…

Strategies for Using the Leveled Texts (cont.)

Above-Grade-Level Students (cont.)

Student-Directed Learning

Because they are academically advanced, above-grade-level students are often the leaders in classrooms. They are more self-sufficient learners, too. As a result, there are some student-directed strategies that teachers can employ successfully with these students. Remember to use the texts in this book as jump starts so that students will be interested in finding out more about the mathematical concepts presented. Above-grade-level students may enjoy any of the following activities:

- Writing their own questions, exchanging their questions with others, and grading the responses.
- Reviewing the lesson and teaching the topic to another group of students.
- Reading other nonfiction texts about these mathematical concepts to further expand their knowledge.
- Writing the quizzes and tests to go along with the texts.
- Creating illustrated timelines to be displayed as visuals for the entire class.
- Putting together multimedia presentations about the mathematical concepts.

Tiered Assignments

Teachers can differentiate lessons by using tiered assignments, or scaffolded lessons. Tiered assignments are parallel tasks designed to have varied levels of depth, complexity, and abstractness. All students work toward one goal, concept, or outcome, but the lesson is tiered to allow for different levels of readiness and performance. As students work, they build on their prior knowledge and understanding. Students are motivated to be successful according to their own readiness and learning preferences.

Guidelines for writing tiered lessons include the following:

1. Pick the skill, concept, or generalization that needs to be learned.
2. Think of an on-grade-level activity that teaches this skill, concept, or generalization.
3. Assess the students using classroom discussions, quizzes, tests, or journal entries and place them in groups.
4. Take another look at the activity from Step 2. Modify this activity to meet the needs of the below-grade-level and above-grade-level learners in the class. Add complexity and depth for the above-grade-level students. Add vocabulary support and concrete examples for the below-grade-level students.

How to Use This Product

Readability Chart

Title of the Text	Star	Circle	Square	Triangle
Measuring the Length of Objects	1.8	3.5	5.0	6.5
Measuring the Weight of Objects	2.2	3.4	5.0	6.5
Measuring Time	2.2	3.4	5.1	6.6
Measuring Temperature	2.3	3.5	5.5	6.5
Measuring Angles	2.2	3.2	5.4	6.7
Estimating Measurements	2.2	3.4	5.4	6.6
Converting Length	2.2	3.5	5.5	6.9
Converting Weight	2.2	3.5	5.5	7.1
Measuring the Perimeter of Regular Shapes	2.0	3.0	5.0	6.5
Measuring the Area of Regular Shapes	2.2	3.5	5.1	6.5
Measuring the Perimeter of Irregular Shapes	2.2	3.0	5.1	6.5
Measuring the Area of Irregular Shapes	2.1	3.5	5.2	6.5
Measuring the Volume of Solids and Liquids	2.1	3.5	5.2	6.6
Converting Volume	2.2	3.3	5.5	6.8
Measuring Surface Area	2.2	3.4	5.0	6.5

Components of the Product

Strong Image Support

- Each level of text includes multiple primary sources. These documents, photographs, and illustrations add interest to the texts. The images also serve as visual support for second-language learners. They make the texts more context-rich and bring the examples to life.

#50754—Leveled Texts for Mathematics: Measurement   © Shell Education

How to Use This Product (cont.)

Components of the Product (cont.)

Practice Problems

- The introduction often includes a challenging question or riddle. The answer can be found on the next page at the end of the lesson.

- Each level of text includes a You Try It section where the students are asked to solve problems using the skill or concept discussed in the text.

- Although the mathematics is the same, the questions may be worded slightly differently depending on the reading level of the passage.

The Levels

- There are 15 topics in this book. Each topic is leveled to four different reading levels. The images and fonts used for each level within a topic look the same.

- Behind each page number, you'll see a shape. These shapes indicate the reading levels of each text so that you can make sure students are working with the correct texts. The reading levels fall into the ranges indicated below. See the chart on page 16 for the specific reading levels of each lesson.

Leveling Process

- The texts in this series were originally authored by mathematics educators. A reading expert went through the texts and leveled each one to create four distinct reading levels.

- A mathematics expert then reviewed each passage for accuracy and mathematical language.

- The texts were then leveled one final time to ensure the editorial changes made during the process kept them within the ranges described to the left.

Levels
1.5–2.2

Levels
3.0–3.5

Levels
5.0–5.5

Levels
6.5–7.2

#50754—*Leveled Texts for Mathematics: Measurement*

How to Use This Product (cont.)

Tips for Managing the Product

How to Prepare the Texts

- When you copy these texts, be sure you set your copier to copy photographs. Run a few test pages and adjust the contrast as necessary. If you want the students to be able to appreciate the images, you need to carefully prepare the texts for them.

- You also have full-color versions of the texts provided in PDF form on the CD. (See page 142 for more information.) Depending on how many copies you need to make, printing the full-color versions and copying those might work best for you.

- Keep in mind that you should copy two-sided to two-sided if you pull the pages out of the book. The shapes behind the page numbers will help you keep the pages organized as you prepare them.

Distributing the Texts

Some teachers wonder about how to hand the texts out within one classroom. They worry that students will feel insulted if they do not get the same papers as their neighbors. The first step in dealing with these texts is to set up your classroom as a place where all students learn at their individual instructional levels. Making this clear as a fact of life in your classroom is key. Otherwise, the students may constantly ask about why their work is different. You do not need to get into the technicalities of the reading levels. Just state it as a fact that every student will not be working on the same assignment every day. If you do this, then passing out the varied levels is not a problem. Just pass them to the correct students as you circle the room.

If you would rather not have students openly aware of the differences in the texts, you can try these ways to pass out the materials:

- Make a pile in your hands from star to triangle. Put your finger between the circle and square levels. As you approach each student, you pull from the top (star), above your finger (circle), below your finger (square), or the bottom (triangle). If you do not hesitate too much in front of each desk, the students will probably not notice.

- Begin the class period with an opening activity. Put the texts in different places around the room. As students work quietly, circulate and direct students to the right locations for retrieving the texts you want them to use.

- Organize the texts in small piles by seating arrangement so that when you arrive at a group of desks you have just the levels you need.

How to Use This Product *(cont.)*

Correlation to Mathematics Standards

Shell Education is committed to producing educational materials that are research and standards based. In this effort, we have correlated all of our products to the academic standards of all 50 United States, the District of Columbia, the Department of Defense Dependent Schools, and all Canadian provinces. We have also correlated to the Common Core State Standards.

How to Find Standards Correlations

To print a customized correlation report of this product for your state, visit our website at **http://www.shelleducation.com** and follow the on-screen directions. If you require assistance in printing correlation reports, please contact Customer Service at 1-877-777-3450.

Purpose and Intent of Standards

Legislation mandates that all states adopt academic standards that identify the skills students will learn in kindergarten through grade twelve. Many states also have standards for Pre-K. This same legislation sets requirements to ensure the standards are detailed and comprehensive.

Standards are designed to focus instruction and guide adoption of curricula. Standards are statements that describe the criteria necessary for students to meet specific academic goals. They define the knowledge, skills, and content students should acquire at each level. Standards are also used to develop standardized tests to evaluate students' academic progress. Teachers are required to demonstrate how their lessons meet state standards. State standards are used in the development of all of our products, so educators can be assured they meet the academic requirements of each state.

TESOL Standards

The lessons in this book promote English language development for English language learners. The standards listed on the Teacher Resource CD support the language objectives presented throughout the lessons.

NCTM Standards Correlation Chart

The chart on the next page shows the correlation to the National Council for Teachers of Mathematics (NCTM) standards. This chart is also available on the Teacher Resource CD (*nctm.pdf*).

NCTM Standards

NCTM Standard	Lesson	Page
Understand such attributes as length, area, weight, volume, and size of angle and select the appropriate type of unit for measuring each attribute	Measuring the Length of Objects; Measuring the Weight of Objects; Measuring Angles; Estimating Measurements; Converting Length; Converting Weight; Measuring the Area of Regular Shapes; Measuring the Area of Irregular Shapes; Measuring the Volume of Solids and Liquids; Converting Volume; Measuring Surface Area	21–29, 53–84, 93–100, 109–140
Understand the need for measuring with standard units and become familiar with standard units in the customary and metric systems	Converting Length; Converting Weight; Converting Volume	69–84, 125–132
Carry out simple unit conversions, such as from centimeters to meters, within a system of measurement	Converting Length; Converting Weight; Converting Volume	69–84, 125–132
Understand that measurements are approximations and how differences in units affect precision	Estimating Measurements	61–68
Explore what happens to measurements of a two-dimensional shape such as its perimeter and area when the shape is changed in some way	Measuring the Perimeter of Regular Shapes; Measuring the Area of Regular Shapes; Measuring the Perimeter of Irregular Shapes; Measuring the Area of Irregular Shapes	85–116
Develop strategies for estimating the perimeters, areas, and volumes of irregular shapes	Estimating Measurements; Measuring the Perimeter of Regular Shapes; Measuring the Perimeter of Irregular Shapes; Measuring the Volume of Solids and Liquids	61–68, 85–92, 101–108, 117–124
Select and apply appropriate standard units and tools to measure length, area, volume, weight, time, temperature, and the size of angles	All lessons	21–140
Select and use benchmarks to estimate measurements	Estimating Measurements	61–68
Develop, understand, and use formulas to find the area of rectangles and related triangles and parallelograms	Measuring the Area of Regular Shapes; Measuring the Area of Irregular Shapes	93–100, 109–116
Develop strategies to determine the surface areas and volumes of rectangular solids	Measuring the Volume of Solids and Liquids; Converting Volume; Measuring Surface Area	117–140

Standards are listed with the permission of the National Council of Teachers of Mathematics (NCTM). NCTM does not endorse the content or validity of these alignments.

Measuring the Length of Objects

Is your foot a foot long? How many inches is your hand? How many millimeters is a paper clip? There are many ways to measure length.

Basic Facts

Length is the measure of a thing from one end to the other. There are many tools used to measure length. But, most of the time, we use a ruler. Yardsticks can be used, too. And, so can meter sticks. Many people use tape measures.

Length Units of Measurement

There are many units you can use to measure length. In the United States, we use English units. But, many other countries do not. They use the metric system.

English Units	Metric Units
inches	millimeters
feet	centimeters
yards	meters
miles	kilometers

So, How Long Is It?

How long is an inch? An inch is about as wide as two fingers. Some say the inch was the width of a king's thumb. But we do not know the true story. An inchworm is about an inch long. So is a small paper clip. One inch is 2.54 centimeters long.

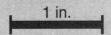

1 in.

A foot is 12 inches. Or it is 30.48 centimeters. A ruler is a foot long. You can even get a hot dog that is a foot long! One story says that the foot was the length of a king's foot.

21

So, How Long Is It? (cont.)

A yard is a measure of length. A yard is three feet. It is 36 inches. Many say that Henry I of England made the yard. It was the length from the tip of his nose to the tip of his finger. This was when his arms were out. It is easy to measure long things with a yardstick. A football field is measured in yards. It is 100 yards long. A yard is very close to a meter.

A mile is a big unit. It is a measure of length. A mile is 5,280 feet. The Romans were the first to use a mile. They used their steps to count. Their steps counted out 5,000 feet. That made their mile. When you go on a trip, you measure with miles. Running tracks are a quarter-mile long. A mile is close to a kilometer. One mile is 1.6 kilometers.

How to Measure Length

The ruler is a tool. It measures things. The ruler is 12 inches. That is one foot. But using a ruler can be tricky. You should pay close attention to the marks on the ruler. You do not start measuring at the start of the ruler. You should begin your measure at the first mark. The arrow below shows where to start measuring on this ruler.

Measuring Length in Our Daily Lives

We measure in many ways. Stores measure cloth in yards. Airlines use miles to measure the distance of their flights. Construction workers use tape measures when building homes. They use them to put in carpet, too. They measure the carpet in feet. There are many ways we measure length every day!

You Try It

Which units would you use to measure these things?

- the length of your desk
- the length of one wall in your classroom
- the length of the hallway
- the length from your house to school

Now choose one of the first three things above.
Measure the length using the correct unit of measurement.

Measuring the Length of Objects

Is your foot really a foot long? How many inches is your hand? How many millimeters is a paper clip? There are many ways to measure the length of things.

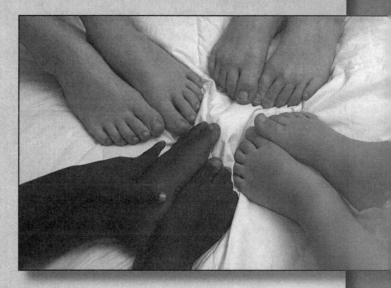

Basic Facts

Length is the measure of a thing from one end to the other end. There are many tools used to measure length. But most of the time we use a ruler. Yardsticks can be used, too. And so can meter sticks. Many people even use tape measures to measure length.

Length Units of Measurement

There are a lot of units you can use to measure length. In the United States, we use English units of measurement. But, in many other countries, they use the metric system.

English Units
inches
feet
yards
miles

Metric Units
millimeters
centimeters
meters
kilometers

So, How Long Is It?

How long is an inch? An inch is about as wide as two fingers. Some say the inch was the width of a king's thumb. But we do not know the true story. An inchworm is about an inch long. So is a small paper clip. One inch is 2.54 centimeters.

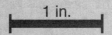

1 in.

A foot is 12 inches. Or it is 30.48 centimeters. A ruler is a foot long. You can even get a hot dog that is a foot long! Legend says that the foot started as the length of a king's foot.

(23)

So, How Long Is It? (cont.)

A yard is used to measure length. A yard is three feet. It is 36 inches. Many say that Henry I of England made the yard. It was the length from the tip of his nose to the tip of his finger when his arms were out. It is easier to measure long things with a yardstick. A football field is measured in yards. It is 100 yards long. A yard is very close to a metric meter.

A mile is a large unit used to measure length. A mile is 5,280 feet. The Romans were the first to use a mile. They measured it using steps. Their steps equaled 5,000 feet. That made their mile. When you go on a trip, you measure the distance in miles. Running tracks are a quarter-mile long. A mile is close to a kilometer. One mile is 1.6 kilometers.

How to Measure Length

The ruler is the tool we use the most to measure things. The ruler measures 12 inches, or one foot. But, using a ruler can be a little tricky. You need to pay close attention to the marks on the ruler. You do not always start measuring at the beginning of the ruler. Instead, you should begin your measurement at the first mark. The arrow below shows where to begin measuring on a standard ruler.

Measuring Length in Our Daily Lives

We use measurements in many ways. Fabric stores measure cloth using yards, airlines use miles to measure the distance of their flights, and construction workers must use tape measures when building homes or installing carpet. Construction workers measure the carpet in feet. There are many ways we measure length every day!

You Try It

Which units of measurement would you use to measure these things?

- the length of your desk

- the length of one wall in your classroom

- the length of the hallway

- the length from your house to school

  Now choose one of the first three things above. Measure the length using the correct unit of measurement.

(24)

Measuring the Length of Objects

Is your foot really a foot long? How many inches is your hand? How many millimeters is a paper clip? There are many ways to measure the length of objects.

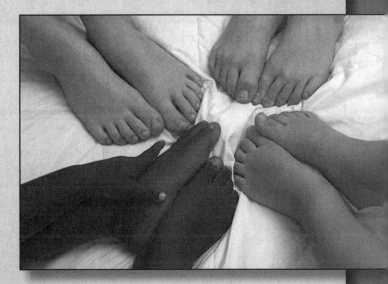

Basic Facts

Length is the measurement of an object from one end to the other end. There are many tools we can use to measure length, but most of the time, we measure short lengths using a ruler. Yardsticks and meter sticks are also often used to measure length. Many people even use tape measures to measure the length of objects.

Length Units of Measurement

There are a lot of units you can use when measuring objects. In the United States, we use English units of measurement. However, in many other countries around the world, the metric system is used.

English Units	Metric Units
inches	millimeters
feet	centimeters
yards	meters
miles	kilometers

So, How Long Is It?

How long is an inch? An inch is about the width of your two fingers. It is said that the inch began as the width of a king's thumb. However, the true history of the inch is not known. An inchworm is about an inch long and so is a small paper clip. One inch equals about 2.54 centimeters.

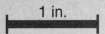

1 in.

A foot equals 12 inches or about 30.48 centimeters. A ruler is one foot long. Sometimes, you can even get a hot dog that is a foot long! Legend also says that the foot started as the length of a king's foot.

25

So, How Long Is It? *(cont.)*

A yard is also used to measure length. A yard is three feet or 36 inches. Many claim that Henry I of England created the yard. It was the length from the tip of his nose to the tip of his finger when his arms were outstretched. It is often easier to measure longer objects using a yardstick instead of a ruler. A football field is measured in yards. It is 100 yards long. A yard is very close to a meter in the metric system.

A mile is one of the largest units used to measure length. A mile is 5,280 feet. The Romans were the first to measure a mile. They measured it using steps. Their steps equaled 5,000 feet to make a mile. When you go on a trip, you measure the distance in miles. Running tracks are a quarter-mile long. A mile is close to a kilometer. One mile equals 1.6 kilometers.

How to Measure Length

The ruler is the tool we use the most to measure objects. The ruler measures 12 inches, or one foot; however, using a ruler can be a little tricky. When measuring an object, you need to pay close attention to the marks on the ruler. You do not always start measuring at the beginning of the ruler. Instead, you should begin your measurement at the first mark. The arrow below shows where to begin measuring on a standard ruler.

Measuring Length in Our Daily Lives

We use measurements in many ways. Fabric stores measure cloth using yards, airlines use miles to measure the distance of their flights, and construction workers must use tape measures when building homes or installing carpet. Construction workers measure the carpet in feet. There are many ways we measure length every day!

You Try It

Which units of measurement would you use to measure the following items?

- the length of your desk
- the length of one wall in your classroom
- the length of the hallway
- the length from your house to school

Choose one of the first three items above and measure the length using the correct unit of measurement.

Measuring the Length of Objects

Is your foot actually a foot long? How many inches is your hand? How many millimeters is a paper clip? There are as many ways to measure the length of objects as there are things to compare them against.

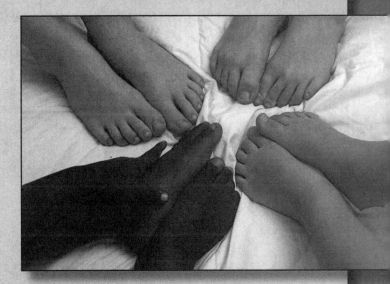

Basic Facts

Length is considered to be the measurement of an object from one end to the other. There are many tools we can use to measure length, but most frequently we measure length using a ruler. Yardsticks, meter sticks, and even tape measures are also often used to determine the length of objects.

Length Units of Measurement

A wide variety of units may be used to measure the length of objects. In the United States, we use the English units of measurement; however, in many other countries around the world, the metric system is primarily used.

English Units	Metric Units
inches	millimeters
feet	centimeters
yards	meters
miles	kilometers

So, How Long Is It?

How long is an inch? An inch is about the width of your two fingers. It is said that the inch began as the width of a king's thumb; however the true history of this unit of measurement is not known. An inchworm is about an inch long and so is a small paper clip. If you were to convert one inch into metric measurements, you would find it equals about 2.54 centimeters.

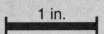

1 in.

A foot equals 12 inches, or about 30.48 centimeters. A classroom ruler is a foot long, and sometimes, you can even get a hot dog that is a foot long! Legend also says that the foot started as the length of a king's foot.

27

So, How Long Is It? *(cont.)*

A yard, which is three feet or 36 inches, is also used to measure length. Many claim that Henry I of England created the yard by measuring the length from the tip of his nose to the tip of his finger when his arms were outstretched. It is often easier to measure longer objects using a yardstick instead of a ruler. For example, a football field is 100 yards long. A yard is very close in length to a meter in the metric system.

A mile, or 5,280 feet, is one of the largest units used to measure length. The Romans were the first to measure a mile by using steps. Their steps equaled 5,000 feet to make a mile. When you go on a trip or run a long race, you measure the distance in miles. A mile can be compared to a kilometer in the metric system: one mile equals 1.6 kilometers.

How to Measure Length

The ruler is the tool we use the most to measure objects. The ruler measures 12 inches, or one foot; however, using a ruler can be a little tricky. When measuring an object, you need to pay close attention to the marks on the ruler because you do not always start measuring at the beginning of the ruler. Instead, you should begin your measurement at the first mark. The arrow below shows where to begin measuring on a standard ruler.

Measuring Length in Our Daily Lives

We use measurements in many ways. Fabric stores measure cloth using yards, airlines use miles to measure the distance of their flights, and construction workers must use tape measures when building homes or installing carpet. Construction workers measure the carpet in feet. There are many ways we measure length every day!

You Try It

Which units of measurement would you use to determine the length of the following items?

- the length of your desk

- the length of one wall in your classroom

- the length of the hallway

- the length from your house to school

Now choose one of the first three items above and measure the length using the correct unit of measurement.

Measuring the Weight of Objects

Does a butterfly weigh more than a hummingbird? Does an elephant or a rhino weigh more?

Animals come in different sizes. They are different weights. So are humans. A big butterfly weighs 0.1 ounce. That is $\frac{1}{10}$ of an ounce. And that is about 3 grams. A big hummingbird weighs 20 grams. That is 0.7 ounce. 0.7 ounce is more than 0.1 ounce. Most elephants weigh more than most rhinos. The average rhino weighs 5,500 pounds. That is over 2 tons! An elephant weighs from 6,000 to 10,000 pounds.

Basic Facts

Weight tells us how heavy a thing is. Scales measure weight. Stores use scales to weigh foods. At home, you may use a different scale to weigh yourself.

Long ago, stones and seeds were used to measure weight. Things were weighed on a balance. Stones were put on one side. Or bags of seeds were put there. The thing was placed on the other side. Today, we use better scales.

Mass and weight are different. On Earth, we measure our weight, which tells the pull of gravity. But on the moon, there is less gravity. So you weigh much less. But your mass is the same on Earth and on the moon.

Weight Units of Measurement

You can use different units to weigh things. In the United States, we use the English units. In many other places they use the metric system.

English Units	Metric Units
ounces	milligrams
pounds	grams
tons	kilograms

How Much Does It Weigh?

What do you use to weigh small things? You should use ounces. A piece of bread weighs about one ounce. So does a group of 10 pennies. An apple weighs about 4 ounces. The post office will weigh a letter in ounces. That is how they know how much it will cost to send it. There are 16 ounces in 1 pound.

We use pounds to weigh heavy things. We use pounds to weigh our bodies. A newborn baby weighs about 7 pounds. A bag of flour weighs about 4 pounds. Our cars weigh about 3,000 pounds.

The ton is a heavy weight. It is the largest unit of measurement for weight. There are 2,000 pounds in one ton. A male hippo weighs almost one ton. Heavy machines in factories can weigh a ton, too.

The ounce, pound, and ton are all English units. A gram is a metric unit. Grams are used to weigh small things. An aspirin weighs a gram. So does a paper clip. Kilograms are used to measure bigger things. A textbook weighs about 1 kilogram.

Measuring Weight in Our Daily Lives

Many people weigh things for their jobs. Think of the doctor's office. Patients are weighed with a scale. Medicines are weighed, too. They are weighed in milligrams. That is a very small amount. It is more precise. A small amount could make the wrong dose of medicine. Grocery stores use weight. Food is weighed with a scale. They use scales at the meat shop. The fish market uses them, too. What other jobs need weight to be measured?

You Try It

Which units would you use to measure the weight of the things below? There can be more than one answer for each thing. Be sure to think about the English units. And think about the metric units.

- a stack of paper clips

- your textbook

- the school bus

Now choose a thing in your classroom. Weigh it with a scale. Be sure to tell which unit of measurement you used.

#50754—Leveled Texts for Mathematics: Measurement

Measuring the Weight of Objects

Does a butterfly weigh more than a hummingbird? Does an elephant or a rhino weigh more?

Animals come in different weights and sizes. So do humans. A larger butterfly weighs 0.1 ounce, or $\frac{1}{10}$ of an ounce. That is about 3 grams. The larger hummingbirds weigh 20 grams, or 0.7 ounce. On average, an elephant weighs more than a rhino. The average rhino weighs about 5,500 pounds. That is over 2 tons! An elephant weighs between 6,000 and 10,000 pounds.

Basic Facts

Weight tells us how heavy something is. Scales measure weight. Stores use scales to weigh foods. At home, you may use a different kind of scale to measure your own weight.

Long ago, stones and seeds were used to measure weight. Things were weighed on a balance. Stones were placed on one side of the balance. The thing was placed on the other side. Today, we use more advanced scales.

Mass and weight are different. On Earth, we measure our weight, which is how much gravity pulls us to Earth. On the moon, gravity is less, so you weigh less. But you still have the same mass on the moon.

Weight Units of Measurement

There are a lot of units you can use when measuring the weight of a thing. In the United States, we use the English units of measurement. But in many other countries around the world, the metric system is used.

English Units	Metric Units
ounces	milligrams
pounds	grams
tons	kilograms

© Shell Education   #50754—Leveled Texts for Mathematics: Measurement

How Much Does It Weigh?

When you measure small things, you should use ounces. A slice of bread weighs about one ounce. So does a group of 10 pennies. An apple weighs about 4 ounces. The post office will weigh a letter in ounces to see how much it will cost to send it. There are 16 ounces in 1 pound.

We use pounds to weigh heavier things. Our bodies are weighed in pounds. A newborn baby weighs about 7 pounds. A bag of flour weighs about 4 pounds. And, our cars weigh about 3,000 pounds.

The ton is the largest unit of measurement for weight. There are 2,000 pounds in one ton. A male hippo weighs almost one ton. Heavy machines in factories can weigh a ton, too.

The ounce, pound, and ton are all part of the English units of measurement. In the metric system, grams are used to weigh small items. An aspirin weighs about a gram. So does a paper clip. Kilograms are used to measure larger items. One kilogram is about the weight of a textbook.

Measuring Weight in Our Daily Lives

Many people weigh things as part of their jobs. At the doctor's office, patients are weighed with a scale. Medicines are weighed, too. They are weighed in milligrams. That is a very small amount. It is more accurate. A small amount could make the wrong dose of medicine. Grocery stores use weight. Food is weighed with a scale. They use scales at the butcher shop. The fish market uses them, too. What other jobs need measurement?

You Try It

Which units would you use to measure the weight of the following things? There can be more than one answer. Be sure to think about both the English units and the metric units.

- a stack of paper clips

- your social studies book

- the school bus

Choose an object in your classroom. Measure its weight using a scale. Tell which unit of measurement you used.

Measuring the Weight of Objects

What weighs more: a butterfly or a hummingbird? An elephant or a rhinoceros?

Animals come in various weights and sizes, just as humans do. A larger butterfly weighs 0.1 ounce, or $\frac{1}{10}$ of an ounce. That equals about 3 grams. The larger hummingbirds weigh 20 grams, or 0.7 ounce. On average, an elephant weighs more than a rhinoceros. The average rhinoceros weighs about 5,500 pounds. That is over two tons! An elephant weighs between 6,000 and 10,000 pounds.

Basic Facts

Weight tells us how heavy something is. Scales are used to measure weight. Grocery stores use scales to measure the weight of foods. At home, you may use a different type of scale to measure your own weight.

Before we used scales with numbers on them, stones and seeds were used to measure weight. The items were weighed on a balance. Stones were placed on one side of the balance and the item was placed on the other side. Today, we use more advanced scales.

Mass and weight are different, but on Earth, they are used the same way. On Earth, weight measures the pull of gravity on your body. On the moon, there is much less gravity, so you actually weigh much less. But even on the moon your mass is the same.

Weight Units of Measurement

There are a lot of units you can use when measuring the weight of an object. In the United States, we use the English units of measurement; however, in many other countries around the world, the metric system is used.

English Units	Metric Units
ounces	milligrams
pounds	grams
tons	kilograms

© Shell Education   #50754—Leveled Texts for Mathematics: Measurement

How Much Does It Weigh?

When you measure small items, you should use ounces. A slice of bread weighs about one ounce and so does a group of 10 pennies. An apple weighs about 4 ounces. When you mail a letter, the post office will weigh it in ounces to determine how much it will cost to send the letter. There are 16 ounces in 1 pound.

We use pounds to weigh heavier items. Our bodies are weighed in pounds. The average newborn baby weighs about 7 pounds. A bag of flour weighs about 4 pounds, and our cars weigh about 3,000 pounds.

The ton is the largest unit of measurement for weight. There are 2,000 pounds in one ton. A male hippopotamus weighs almost one ton. Heavy machinery in factories can weigh a ton, too.

The ounce, pound, and ton are all part of the English units of measurement. In the metric system, grams are used to weigh small items. An aspirin or paper clip weighs about a gram. And kilograms are used to measure larger items. One kilogram is about the weight of a textbook.

Measuring Weight in Our Daily Lives

Many jobs require the weight of items to be measured. At the doctor's office, patients are weighed using a scale. Medicines are weighed, often in milligrams. The milligram is a very small amount, and, therefore, provides more accuracy. A single milligram could make the wrong dose of medicine. Grocery stores also use weight. Food is weighed using a scale at the butcher shop or fish market. What other jobs require people to measure weight?

You Try It

Which units would you use to measure the weight of the following items? There can be more than one answer for each item. Be sure to think about both the English units of measurement and the metric system.

- a stack of paper clips
- your social studies book
- the school bus

Now, choose an object in your classroom and measure the weight of that object using a scale. Be sure to tell which unit of measurement you used.

#50754—*Leveled Texts for Mathematics: Measurement*

Measuring the Weight of Objects

Which animal weighs more: a butterfly or a hummingbird? An elephant or a rhinoceros?

Animals come in various weights and sizes, just as humans do. A larger butterfly weighs 0.1 ounce, or $\frac{1}{10}$ of an ounce, which is the same as about 3 grams, while larger hummingbirds weigh 20 grams, or 0.7 ounce. On average, an elephant weighs more than a rhinoceros, which is especially impressive since the average rhinoceros weighs 5,500 pounds. That is over 2 tons! An adult elephant can weigh anywhere from 6,000 to 10,000 pounds.

Basic Facts

Weight is used to determine how heavy something is. Scales, like those used in grocery stores to weigh foods, are used to measure weight. At home, you may use a different type of scale to measure your own weight.

Before humans developed scales with numbers on them, stones and seeds were used to measure weight with a tool called a balance. Stones were placed on one side of the balance and the item to be weighed was placed on the other side. Today we use more sophisticated scales to measure weight.

Mass and weight are two different things. On Earth, our weight measures gravity pulling on us. But on the moon, there is much less gravity, so you actually weigh less. However, your mass does not change.

Weight Units of Measurement

There are a variety of units you can use when measuring the weight of an object. The United States generally uses the English units of measurement; however, in many other countries around the world, the metric system is the standard for use.

English Units	Metric Units
ounces	milligrams
pounds	grams
tons	kilograms

© Shell Education   #50754—Leveled Texts for Mathematics: Measurement

How Much Does It Weigh?

When you measure small items, you should use ounces. A slice of bread weighs about one ounce as does a group of 10 pennies, while an apple weighs about 4 ounces. When you mail a letter, the post office will weigh it in ounces to determine how much it will cost to send the letter. There are 16 ounces in 1 pound.

We use pounds to weigh heavier items like dry goods, automobiles, or the human body. The average newborn baby weighs 7 pounds, while a bag of flour weighs 4 pounds, and our cars often weigh 3,000 pounds.

The ton, which measures 2,000 pounds, is the largest unit of measurement for weight. A male hippopotamus weighs almost one ton and heavy machinery in factories can weigh a ton or more, as well.

The ounce, pound, and ton are all part of the English units of measurement. In the metric system, grams are used to weigh small items like aspirin or paper clips, each of which weigh about 1 gram. Kilograms are used to measure larger items. For instance, one kilogram is about the weight of a textbook.

Measuring Weight in Our Daily Lives

Many jobs require weight to be measured. At the doctor's office, patients are weighed using a scale. To determine the proper dosage, medicines are also weighed, often in milligrams. The milligram is a very small amount; therefore, the measurements are more accurate. A single milligram could make the wrong dose of medicine. Grocery stores also use weight. Food is weighed using a scale at the butcher shop or fish market. What other jobs require people to measure weight?

You Try It

Which units would you use to measure the weight of the following items? Consider both the English units of measurement and the metric system.

- a stack of paper clips
- the school bus
- your social studies book

Choose an object in your class and measure its weight using a scale. Tell which unit of measurement you used.

Measuring Time

You wake up late today. You have 30 minutes to get to school on time. It takes you 8 minutes to shower. It takes 5 minutes to get dressed. You need 3 minutes to brush your teeth. It takes 9 minutes to eat breakfast. And it takes 10 minutes to get to school. Will you be on time? Or, will you be late this morning?

Basic Facts

How do we measure time? We use seconds. We use minutes. And we use hours. There are 60 seconds in a minute. There are 60 minutes in an hour. There are 3,600 seconds in an hour. Can you figure how many seconds are in a day?

Days are parts of time. Weeks are parts of time. Years are, too. A year is 12 months. A decade is 10 years. A century is 100 years. And 1,000 years makes a millennium. Can you figure how many months are in a century? How many months have you lived?

Telling time begins by knowing about clocks. The small hand shows the hours. The large hand shows the minutes. The large hash marks stand for five minutes. The small hash marks stand for one minute.

Objects Used in Telling Time: Past and Present

Today we have things that help us to easily tell time. Most classrooms have a clock on the wall. Many people wear watches. Cell phones show the time. MP3 players give the time, too. Stopwatches measure the amount of time it takes to finish a task. There are so many helpful items to give the time. It should be almost impossible to be late!

Telling time has not always been easy. Long ago, people used the sun to tell time. And they used the moon. They used the stars to help them. And they used the planets, too. They even used major events to measure time. In ancient Egypt, the Nile River would flood. This was used to tell when a year had passed. This is because the Nile flooded about every 365 days. The moon's cycle showed a month. This is because the moon's cycle takes about 29.5 days. And the rising and setting of the sun helped show when one day had passed.

Objects Used in Telling Time: Past and Present (cont.)

Even shadows helped to tell time. The ancient Egyptians would put a tall stick in the ground. The stick's shadow would move as the sun moved across the sky. This would show how much time had passed in the day.

Sundials were like shadow sticks. But, they had a flat top. On this top were numbers. A dial on the top cast a shadow. This showed the time of day. But, sundials could not work at night. So, water clocks were made. Water dripped from one part of the clock to the other. The water level told the time. An hourglass is like this. But, it lets sand fall to tell a given amount of time. Can you think of other things that help us tell time?

Elapsed Time

Hourglasses are a tool. They can tell us how much time has passed. This is called **elapsed time**. Stopwatches show this, too. So do shadow sticks. Knowing how much time has passed is important. It is important for sports matches. Officials must know how many minutes have passed. This is so that they do not let the match run too long. In school, teachers must know how much time has passed. This is so they will not be late to begin the next class. Being able to tell elapsed time helps us keep track of the time. It helps us to be where we need to be.

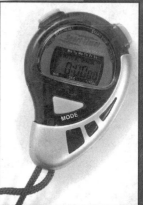

On the clocks (left), can you tell how much time has passed? The first clock reads 1:10. The second clock's time is 3:27. This means that two hours and 17 minutes has elapsed between the times.

Measuring Time in Our Daily Lives

Telling time begins at a young age. Young children are given minutes in time-out. At school, students must know what time it is for their daily schedules. In sports, quarters and halves are measured in time. Time is all around us. Knowing how to tell time is important.

You Try It

Answer these questions about telling time.

- How much time passes from the moment you begin school until the end of the school day?

- How much time elapses between the time you get home to the time you go back to school the next morning?

Measuring Time

You wake up late this morning. You have 30 minutes to get to school on time. It takes you 8 minutes to shower. It takes 5 minutes to get dressed. You need 3 minutes to brush your teeth. It takes 9 minutes to eat breakfast. And it takes 10 minutes to get to school. Will you be on time or late this morning?

Basic Facts

How do we measure time? We use seconds, minutes, and hours. There are 60 seconds in a minute. There are 60 minutes in an hour. There are 3,600 seconds in an hour. Can you tell how many seconds are in a day?

Days, weeks, months, and years are all parts of time. A year is 12 months. A decade is 10 years. A century is 100 years. And 1,000 years makes a millennium. Can you figure how many months are in a century? How many months have you lived?

Telling time begins by understanding the basics about clocks. It is important to understand that the small hand on a clock shows the hours and the large hand shows the minutes. The large hash marks on the clock represent five minutes. The small hash marks on the clock represent one minute.

Objects Used in Telling Time: Past and Present

Today we have things that help us to easily tell time. Most classrooms have a clock hanging on the wall. Many people wear watches to tell the time. Cell phones show the time. MP3 players and other electronic devices give the time, as well. Stopwatches help measure the amount of time it takes to finish a task. With all of the helpful items to give the time, it should be almost impossible to be late!

But telling time has not always been that easy. Long ago, people used the position of the sun and moon to tell time. They used the stars and planets to help them tell the time, too. They even used major events to measure time. In ancient Egypt, the flooding of the Nile River was used to tell when a year had passed. This is because the Nile flooded about every 365 days. The moon's cycle showed a month. This is because the moon's cycle takes about 29.5 days. And the rising and setting of the sun helped show when one day had passed.

(39)

Objects Used in Telling Time: Past and Present (cont.)

Even shadows helped to tell time. The ancient Egyptians would put a tall stick in the ground. The stick's shadow would move as the sun moved across the sky. This would show how much time had passed in the day.

Sundials were like shadow sticks, but they had a flat top. On this top were numbers. A dial on the top cast a shadow. This showed the time of day. But sundials could not work at night. So water clocks were created. Water dripped from one part of the clock to the other. The water level told the time. This is like an hourglass, which lets sand fall to tell a given amount of time. Can you think of other things that help us tell time?

Elapsed Time

Hourglasses, shadow sticks, and stopwatches can tell us how much time has passed. This is called **elapsed time**. Knowing the elapsed time is important. In sports matches, umpires and referees must know how many minutes have passed. This is so that they do not let the match run too long. In school, teachers must know how much time has elapsed since class started. This is so they will not be late to begin the next class. Being able to tell elapsed time helps us keep track of the time. It helps us to stay on schedule.

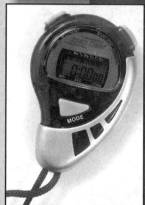

On the clocks (left), can you tell how much time has elapsed? The first clock reads 1:10. The second clock's time is 3:27. This means that two hours and 17 minutes has elapsed between the times on the two clocks.

Measuring Time in Our Daily Lives

Telling time begins at a young age. Young children are given minutes in time-out, and at school students must know what time it is for their daily schedules. At sporting events, quarters and halves are measured in time. Time is all around us and knowing how to tell time is important.

You Try It

Answer the following questions about telling time.

- How much time elapses from the moment you begin school until the end of the school day?
- How much time elapses between the time you get home to the time you go back to school the next morning?

Measuring Time

You accidentally wake up late in the morning. You have 30 minutes to get to school on time. It takes you 8 minutes to shower, 5 minutes to get dressed, 3 minutes to brush your teeth, 9 minutes to eat breakfast, and 10 minutes to get to school. Will you be on time or late?

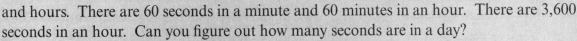

Basic Facts

We measure amounts of time in seconds, minutes, and hours. There are 60 seconds in a minute and 60 minutes in an hour. There are 3,600 seconds in an hour. Can you figure out how many seconds are in a day?

Days, weeks, months, and years are also elements of time. A year is 12 months, and a decade is 10 years. A century is 100 years, and 1,000 years makes a millennium. Can you figure out how many months are in a century? How many months have you lived?

Telling time begins by understanding the basics about clocks. It is important to understand that the small hand on a clock shows the hours and the large hand shows the minutes. The large hash marks on the clock represent five minutes. The small hash marks on the clock represent one minute.

Objects Used in Telling Time: Past and Present

Today we have objects that help us to easily tell time. Most classrooms have a clock hanging on the wall. Many people wear watches to tell the time. Cell phones, MP3 players, and other electronic devices give the time. Stopwatches are used to measure the amount of time it takes to complete a task. With all of the updated devices to give the time, it should be almost impossible to be late!

However, telling time has not always been that easy. Long ago, people used the position of the sun and moon to tell time. They also depended on the stars and planets to help them tell the time. Major events were also used to measure time. In ancient Egypt, the flooding of the Nile River was used to tell when a year had passed. This is because the Nile flooded about every 365 days. A month was determined by the moon's cycle because the moon's cycle takes about 29.5 days. And the rising and setting of the sun helped determine when one day had passed.

41

Objects Used in Telling Time: Past and Present (cont.)

Shadows also helped to tell time. To tell how much time had elapsed during a day, the ancient Egyptians placed a tall stick in the ground. The stick's shadow would move as the sun moved across the sky. This would show how much time had passed in the day.

Sundials were similar to shadow sticks, but they had a flat top with numbers. The dial on the top cast a shadow, showing the time of day. Since sundials could not work at night, water clocks were created. Water dripped from one part of the clock to the other and the water level told the time. This is similar to an hourglass, which lets sand fall to tell a given amount of time. Can you think of other devices that help us tell time?

Elapsed Time

Hourglasses, shadow sticks, and stopwatches can tell us how much time has passed. Knowing the **elapsed time** is important. In sports matches, umpires and referees must know how many minutes have elapsed so that they do not let the match run too long. In school, teachers must know how much time has elapsed since class started so as not to be late to begin the next class. Being able to tell elapsed time helps us keep track of the time and stay on schedule.

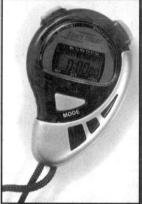

On the clocks (left), can you tell how much time has elapsed? The first clock reads 1:10. The second clock's time is 3:27. This means that two hours and 17 minutes has elapsed between the times on the two clocks.

Measuring Time in Our Daily Lives

Telling time begins at a young age. Young children are given minutes in time-out, and at school students must know what time it is for their daily schedules. At sporting events, play periods are measured in quarters and halves. Time is all around us and knowing how to tell time is important.

You Try It

Answer the following questions about telling time.

- How much time elapses from the moment you begin school until the end of the school day?

- How much time elapses between the time you get home to the time you go back to school the next morning?

Measuring Time

You accidentally wake up late. You have 30 minutes to get to school on time, but it takes you 8 minutes to shower, 5 minutes to get dressed, 3 minutes to brush your teeth, 9 minutes to eat breakfast, and 10 minutes to get to school. Will you be on time or late to school?

Basic Facts

We measure amounts of time in seconds, minutes, and hours. There are 60 seconds in a minute and 60 minutes in an hour, thus there are 3,600 seconds in an hour. Can you figure out how many seconds are in a day?

Days, weeks, months, and years are also elements of time. A year is 12 months, a decade is 10 years, a century is 100 years, and a millennium consists of 1,000 years. Can you figure out how many months are in a century? How many months have you been alive?

Telling time begins by understanding the basics about clocks. It is important to understand that the small hand on a clock shows the hours and the large hand shows the minutes. The large hash marks on the clock represent five minutes, while the small hash marks on the clock each represent one minute.

Objects Used in Telling Time: Past and Present

Today we have objects that help us to easily determine time no matter where we may be. Most classrooms have a clock hanging on the wall; many people wear watches; and cell phones, MP3 players, and other electronic devices display time for our convenience. Stopwatches measure the amount of time it takes to complete a particular task, so with all of the updated devices to give the time, it should be almost impossible to be late!

However, telling time has not always been so convenient. Long ago, people used the position of celestial bodies to keep track of the passage of time. They depended on the sun, the moon, the stars, and planets to help them know the time of day, and the days of the year. Major events were also used to measure time. For example, in ancient Egypt, the flooding of the Nile River was so predictable that it was used to tell when a year had passed, because the Nile flooded about every 365 days. A month was determined by the moon's cycle which takes about 29.5 days. And the rising and setting of the sun helped determine when one day had passed.

43

Objects Used in Telling Time: Past and Present (cont.)

Shadows from the sun can also help to determine time of day. To determine how much time had elapsed during a day, the ancient Egyptians placed a tall stick in the ground. The stick's shadow would move as the sun moved across the sky, showing how much time had elapsed since sunrise.

Sundials were similar to shadow sticks but they had a flat top with numbers inscribed on it. The dial on the top cast a shadow that fell on the numbers, showing the time of day. Since sundials could not work at night, water clocks were created. Water dripped from one part of the clock to the other and the water level told the time. This is similar to an hourglass, which lets sand fall to tell a given amount of time. Can you think of other devices that help us tell time?

Elapsed Time

Hourglasses, shadow sticks, and stopwatches can tell us how much time has passed. Knowing the **elapsed time** is important. For instance, in sports matches, umpires and referees must know how many minutes have elapsed so that they do not let the match run too long, and in school, teachers must know how much time has elapsed since class started so as not to be late to begin the next class. Being able to tell elapsed time helps us keep track of the time and stay on schedule.

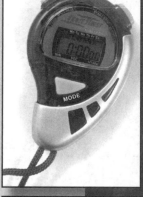

On the clocks (left), can you tell how much time has elapsed? The first clock reads 1:10 and the second clock's time is 3:27. This means that two hours and 17 minutes has elapsed between the times on the two clocks at left.

Measuring Time in Our Daily Lives

Telling time usually begins at a young age. Young children are given minutes in time-out, and at school students must know what time it is for their daily schedules. At sporting events, quarters and halves are measured in time. Time is all around us and knowing how to tell time is important.

You Try It

Answer the following questions about telling time.

- How much time elapses from the moment you begin school until the end of the school day?

- How much time elapses between the time you get home to the time you go back to school the next morning?

Measuring Temperature

You wake up. Your temperature is 101.7 degrees Fahrenheit. Should you go to school? What if your temperature is 98.7 degrees? Our bodies have their own temperatures. They go up. And they go down. They change when we are sick. They go up when we get hot. They go down when we get cold. This is one way we use temperature every day. What are some others?

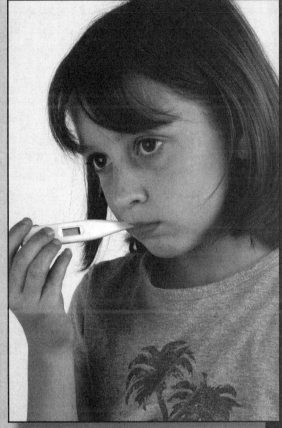

Measuring Temperature

We use a thermometer to measure temperature. There are different kinds. Some take our body's temperature. Other thermometers find the outdoor temperature. We use thermometers to cook. You need the right temperature to cook. You don't want to burn the food. Most homes have thermostats. These control the temperature. And they help you change it.

People in the United States measure temperature. They use the Fahrenheit scale. There is a metric scale, too. It uses Celsius. Most countries use that scale. China does. France does. Canada does, too.

Daniel Fahrenheit made a temperature scale. It is called the Fahrenheit scale. And he made a thermometer to use it. He did this in 1714. On his scale, water freezes at 32 degrees. It boils at 212 degrees. These points are 180 degrees apart.

Andres Celsius made the Celsius scale. His scale is different. Water freezes at 0 degrees. It boils at 100 degrees. These points are 100 degrees apart. 100 is easy to work with.

These men made thermometers. But, they were not the first. The first thermometer was made long ago. It was called a *thermoscope*. An Italian inventor put numbers on it. This was to measure temperature. He was named Santorio Santorio. And Galileo Galilei made an early water thermometer. He did this in 1593. This showed the changes in temperature.

How Does a Thermometer Work?

The bulb thermometer is filled with fluid. Most use mercury. Mercury changes its volume. It uses more space when it is warm. It takes up less space when it is cold. Think of a thermometer. In hot weather, the mercury goes up. The temperature is higher. In cold weather, it takes up less space. It goes down.

Measuring Temperature on a Thermometer

Look at these two thermometers. Can you measure the temperature? Can you find the degrees Fahrenheit on each thermometer? You must think. You must look carefully. Look at the lines on the scale. The lines are two degrees apart.

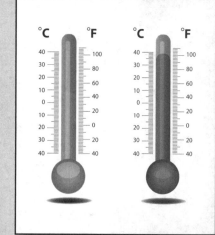

The first thermometer shows 47 degrees Fahrenheit. The second shows 100 degrees Fahrenheit.

Here are two equations. You can use them to change between Fahrenheit and Celsius. The first is to change Fahrenheit to Celsius. Use $C = (F - 32) \times \frac{5}{9}$. The second is to change Celsius to Fahrenheit. Use $F = (\frac{9}{5} \times C) + 32$.

The temperature is 42 degrees Fahrenheit. You want to know Celsius. You should:

1. Subtract 42 – 32. This gives you 10.

2. Multiply your answer by $\frac{5}{9}$. $10 \times \frac{5}{9} = 5.56$ degrees Celsius.

The temperature is 7 degrees Celsius. You want to know Fahrenheit. You should:

1. Multiply 7 by $\frac{9}{5}$. This gives you 12.6.

2. Add 32 to your answer. $12.6 + 32 = 44.6$ degrees Fahrenheit.

Measuring Temperature in Our Daily Lives

Many jobs use temperature. Meteorologists study weather. They find the temperatures of the air and ground. This lets them know if it will be sunny or stormy. They use temperatures to help us plan. Oceanographers are scientists. They study oceans. They need to know the temperatures of the water. They check it on top. They check the temperature deep down. This helps them to learn.

You Try It

Look at the thermometer shown here. Measure the temperature. What is the temperature in Celsius?

#50754—*Leveled Texts for Mathematics: Measurement*

Measuring Temperature

You wake up in the morning. You have a temperature of 101.7 degrees Fahrenheit. Would you have to go to school? What if your temperature is 98.7 degrees? Our bodies have their own temperatures. They rise and fall when we are sick. They change when we get hot or cold. What other ways do we use temperature in our daily lives?

Measuring Temperature

We use a thermometer to measure temperature. There are different kinds of thermometers. Some are made to take our body's temperature. Other thermometers find the outdoor temperature. Most homes have thermostats to control their temperature.

People in the United States measure temperature. They use the Fahrenheit scale. There is also a metric scale. For that, temperature is measured using degrees Celsius. Many countries around the world use metric. China does.

Daniel Fahrenheit made the Fahrenheit scale. And he made a thermometer to show this. His thermometer uses mercury. He did this in 1714. On his scale, the freezing and boiling points of water are separated by 180 degrees. Water freezes at 32 degrees. It boils at 212 degrees.

Andres Celsius made the Celsius scale. His scale is different. The freezing and boiling points of water are 100 degrees apart. Water freezes at 0 degrees. It boils at 100 degrees.

These men were not the first to invent thermometers. The first thermometer was called a *thermoscope*. An Italian inventor put numbers on it. This was to measure temperature. He was named Santorio Santorio. And Galileo Galilei made the early water thermometer. He did this in 1593. This showed the changes in temperature.

How Does a Thermometer Work?

The bulb thermometer is the most common. This thermometer is filled with fluid. Most use mercury. Liquid changes its volume with the temperature. It will take up more space when it is warm. It will take less when it is cold. In hot weather, the mercury rises. In colder weather, it takes up less space. This makes it fall on the thermometer's scale.

Measuring Temperature on a Thermometer

Look at the two thermometers to the right. Can you measure the temperature in degrees Fahrenheit on each thermometer? You must pay close attention. Look at the lines on the scale. The lines are two degrees apart.

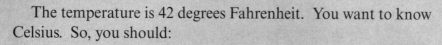

The first thermometer shows 47 degrees Fahrenheit. The second shows 100 degrees Fahrenheit.

Here are two equations. You can use them to change between Fahrenheit and Celsius. The first is to change Fahrenheit to Celsius. Use $C = (F - 32) \times \frac{5}{9}$. The second is to change Celsius to Fahrenheit. Use $F = (\frac{9}{5} \times C) + 32$. Below are examples of how to use the equations to convert temperature.

The temperature is 42 degrees Fahrenheit. You want to know Celsius. So, you should:

1. Subtract 42 – 32. This gives you 10.

2. Multiply your answer by $\frac{5}{9}$. $10 \times \frac{5}{9} = 5.56$ degrees Celsius.

The temperature is 7 degrees Celsius. You want to know Fahrenheit. You should:

1. Multiply 7 by $\frac{9}{5}$. This gives you 12.6.

2. Add 32 to your answer. So 12.6 + 32 = 44.6 degrees Fahrenheit.

Measuring Temperature in Our Daily Lives

There are many jobs that need to find temperature. Meteorologists study weather. They find the temperatures of the air. This lets them predict our weather. Oceanographers study oceans. They need to know the temperatures of the waters where they study.

You Try It

Measure the temperature on the thermometer (right). What is the temperature in Celsius?

Measuring Temperature

You wake up in the morning. You have a temperature of 101.7 degrees Fahrenheit. Would you have to go to school? What if your temperature is 98.7 degrees? Our bodies have internal temperatures. They rise and fall when we are sick or when we get extremely hot or cold. Besides body temperature, in what other ways do we use temperature in our daily lives?

Measuring Temperature

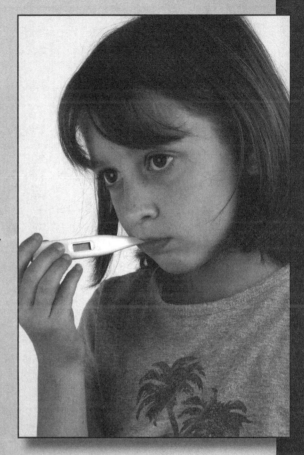

We use a thermometer to measure temperature. There are different types of thermometers. Some of them are made to take our body's temperature. Other thermometers measure the outdoor temperature. In most homes, thermostats are used to regulate the temperature.

In the United States, temperature is measured using the Fahrenheit scale. Temperature on the metric scale is measured using degrees Celsius. Many other countries around the world use this form of measurement, such as China.

Daniel Fahrenheit created the Fahrenheit scale in 1714. On his scale, the freezing and boiling points of water are separated by 180 degrees. Water freezes at 32 degrees. It boils at 212 degrees. Daniel Fahrenheit then created a thermometer using mercury to show this.

Andres Celsius created the Celsius scale. He separated the freezing and boiling points of water by 100 degrees. Water freezes at 0 degrees. It boils at 100 degrees.

However, neither of these men were the first to invent thermometers. The first thermometer was called a *thermoscope*. There was an Italian inventor named Santorio Santorio. He was the first to put numbers on the thermoscope to measure temperature. And Galileo Galilei created the rudimentary water thermometer in 1593. This measured the changes in temperature.

How Does a Thermometer Work?

The bulb thermometer is the most common thermometer we use to measure temperature. This thermometer is filled with mercury. Since liquid changes its volume with the rise and fall of temperatures, it will take up more space when it is warm and less when it is cold. So in hot weather, the mercury rises. And in colder weather, it takes up less space, which causes the mercury to fall on the thermometer's scale.

49

Measuring Temperature on a Thermometer

Look at the two thermometers pictured at right. Can you measure the temperature in degrees Fahrenheit of each thermometer? To measure temperature, you must pay close attention to the lines on the scale. The lines are separated by two degrees.

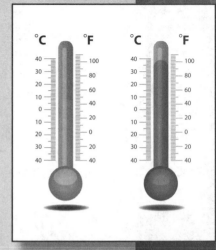

The first thermometer measures 47 degrees Fahrenheit. The second measures 100 degrees Fahrenheit.

There are two equations you can use to convert Celsius to Fahrenheit and Fahrenheit to Celsius. To convert Fahrenheit to Celsius, use $C = (F - 32) \times \frac{5}{9}$. To convert Celsius to Fahrenheit, use $F = (\frac{9}{5} \times C) + 32$. Below are examples of how to use the equations to convert temperature.

If the temperature is 42 degrees Fahrenheit, and you want to convert to Celsius, you should:

1. Subtract $42 - 32$. This gives you 10.

2. Multiply your answer by $\frac{5}{9}$. So, $10 \times \frac{5}{9} = 5.56$ degrees Celsius.

If the temperature is 7 degrees Celsius, and you want to convert to Fahrenheit, you should:

1. Multiply 7 by $\frac{9}{5}$. This gives you 12.6.

2. Add 32 to your answer. So, $12.6 + 32 = 44.6$ degrees Fahrenheit.

Measuring Temperature in Our Daily Lives

There are many jobs that depend on measuring temperature accurately. Meteorologists gauge the temperatures and predict our forecasts. Oceanographers pay close attention to the temperatures of the waters where they explore.

You Try It

Measure the temperature on the thermometer (right). What is the temperature in Celsius?

#50754—Leveled Texts for Mathematics: Measurement

Measuring Temperature

If you wake up in the morning with a temperature of 101.7 degrees Fahrenheit, would you have to go to school? What if your temperature is 98.7 degrees? Our bodies have internal temperatures that rise and fall when we are sick or when we get extremely hot or cold. Besides body temperature, in what other ways do we use temperature in our daily lives?

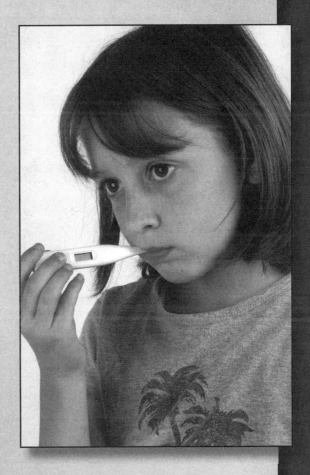

Measuring Temperature

We can use a variety of thermometers to measure temperature. Some of them are made to take our body's temperature, while other thermometers measure the outdoor temperature. In most homes, thermostats are used to regulate and control the temperature.

In the United States, temperature is measured using the Fahrenheit scale. Temperature on the metric scale is measured using degrees Celsius. Many other countries around the world, such as China, use this form of measurement.

Daniel Fahrenheit created the Fahrenheit scale in 1714. On Fahrenheit's scale, the freezing and boiling points of water are separated by 180 degrees; water freezes at 32 degrees and it boils at 212 degrees. Daniel Fahrenheit then created a Length thermometer using mercury to show this.

Andres Celsius created the Celsius scale. He separated the freezing and boiling points of water by 100 degrees, with water freezing at 0 degrees and boiling at 100 degrees.

However, neither of these men were the first to invent thermometers. The first thermometer was called a *thermoscope*. There was an Italian inventor named Santorio Santorio who was the first to put numbers on the thermoscope to measure temperature. And Galileo Galilei created a rudimentary water thermometer in 1593 that measured the changes in temperature.

How Does a Thermometer Work?

The bulb thermometer is the most common thermometer we use to measure temperature. This thermometer is filled with a fluid, often mercury. Since liquid changes its volume with the rise and fall of temperatures, it will take up more space when it is warm and less when it is cold. Thus in hot weather, the mercury rises while in colder weather, it takes up less space, which causes the mercury to fall on the thermometer's scale.

51

Measuring Temperature on a Thermometer

Look at the two thermometers pictured at right. Can you measure the temperature in degrees Fahrenheit of each thermometer? To measure temperature, you must pay close attention to the lines on the scale. The lines are separated by two degrees.

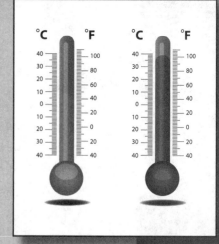

The first thermometer measures 47 degrees Fahrenheit and the second measures 100 degrees Fahrenheit.

There are two equations you can use to convert Celsius to Fahrenheit and Fahrenheit to Celsius. To convert Fahrenheit to Celsius, use $C = (F - 32) \times \frac{5}{9}$. To convert Celsius to Fahrenheit, use $F = (\frac{9}{5} \times C) + 32$. Below are examples of how to use the equations to convert temperature.

If the temperature is 42 degrees Fahrenheit, and you want to convert to Celsius, you should do the following:

1. Subtract $42 - 32$. This gives you 10.

2. Multiply 7 by $\frac{5}{9}$. So, $10 \times \frac{5}{9} = 5.56$ degrees Celsius.

If the temperature is 7 degrees Celsius, and you want to convert to Fahrenheit, you should do the following:

1. Multiply your answer by $\frac{9}{5}$. This gives you 12.6.

2. Add 32 to your answer. So, $12.6 + 32 = 44.6$ degrees Fahrenheit.

Measuring Temperature in Our Daily Lives

Many jobs depend on measuring temperature. Meteorologists gauge temperatures and make forecasts and oceanographers pay close attention to the water temperatures where they explore.

You Try It

Measure the temperature on the thermometer (right). What is the temperature in Celsius?

Measuring Angles

How many angles are in this shape? Are you sure? Did you count them all?

What Are Angles?

An angle is made when two rays meet. The point where they meet is the vertex. The rays are the legs of the angle.	ray angle vertex ray
The size of the angle is measured in degrees. A right angle is 90 degrees. A straight line is 180 degrees. Think of cutting a line in half. Do this by making a vertical line. You will get two 90-degree angles.	180° 90°
An obtuse angle is more than 90 degrees. It is less than 180 degrees.	110°
An acute angle is less than 90 degrees.	70°

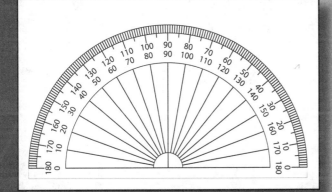

Tools Used to Measure Angles

There are many tools to measure angles. The most common is the **protractor**.

This protractor is a semi-circle. It is used to find the degrees of an angle. It is marked from 0–180. This is for the degrees.

© Shell Education   #50754—Leveled Texts for Mathematics: Measurement

Using a Protractor

We use a protractor to measure angles. This is easy! But you need to use the right steps.

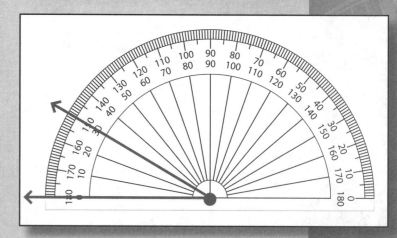

1. First find the endpoint of the rays. This can be called the vertex. Always put it at the center of the protractor.

2. Line the protractor up with one of the rays. Make sure it is even.

3. Look at the second ray. Check the number of degrees that lines up with it. This is the measure of the angle.

What is the measure of the angle above?

Obtuse angles are bigger than 90 degrees. Acute angles are smaller than 90 degrees. Use the right row of numbers. Use the one that would fit that kind of angle. Use big numbers for big angles. Use small numbers for small angles.

Measuring Angles in Our Daily Lives

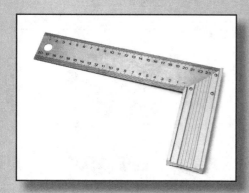

Many people measure angles in their jobs. Architects do this. Engineers do, too. And construction workers do. They must be careful. What would happen if they measured wrong? We would have crooked houses! Our walls might lean. Many wood workers use set squares. And they use miter saws. These help to measure and cut wood at angles. Look at your own home. You will see many angles. They were measured when it was being built!

You Try It

Is each angle obtuse, acute, or right? Now use a protractor to get the exact measurements of the three angles.

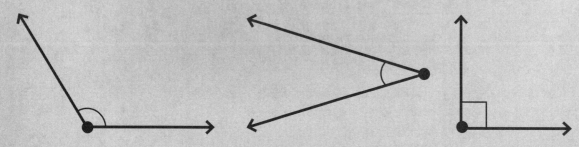

#50754—Leveled Texts for Mathematics: Measurement

Measuring Angles

How many angles are in this shape? Are you sure? Did you count them all?

What Are Angles?

An angle is what is made when two rays meet. The vertex is the point where the rays meet. The two rays are the legs of the angle.	ray / angle / vertex / ray
The size of the angle is measured in degrees. A right angle is 90 degrees. A straight line is 180 degrees. Think of cutting the line in the middle. Do this by making a vertical line. You will get two 90-degree angles.	180° / 90°
An obtuse angle is more than 90 degrees. It is less than 180 degrees.	110°
An acute angle is less than 90 degrees.	70°

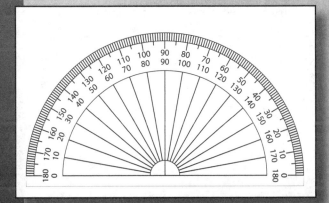

Tools Used to Measure Angles

There are many tools to measure angles. The most common is the **protractor**.

This protractor is a semi-circle. It is marked from 0–180 degrees. It is the tool used most often to find the degrees of an angle.

#50754—Leveled Texts for Mathematics: Measurement

Using a Protractor

We use a protractor to measure angles. This is easy as long as you follow the right steps.

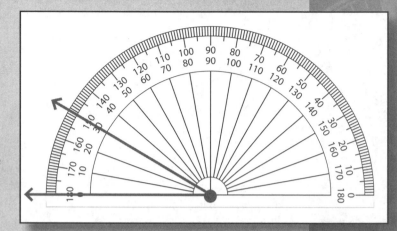

1. Always put the endpoint of the rays, or vertex of the angle on the center of the protractor.

2. Line the protractor up with one of the rays.

3. Look at the number of degrees that lines up with the second ray. This is the measure of the angle.

What is the measurement of the angle pictured above?

Remember, obtuse angles are bigger than 90 degrees. Acute angles are smaller than 90 degrees. Use the row of numbers that would fit that kind of angle.

Measuring Angles in Our Daily Lives

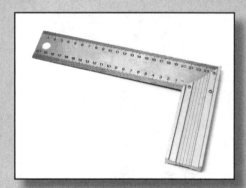

Many people measure angles in their jobs. Architects do this. Engineers do, too. And construction workers do. They must be careful. What would happen if these angles were measured wrong? We would have crooked houses! Many construction workers use set squares and miter saws. These help to measure and cut wood at angles. Look at your own home. You will see the many angles that were measured when it was being built!

You Try It

Decide if each angle is an obtuse, acute, or right angle. Then use a protractor to get the exact measurements of the three angles.

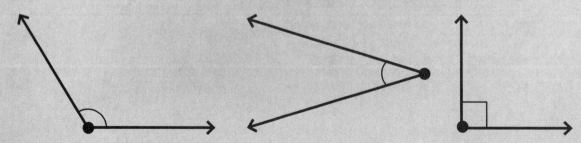

Measuring Angles

How many angles do you see in this shape?
Are you sure you counted them all?

What Are Angles?

An angle is made when two rays meet. When the rays meet, they form a vertex. The two rays form the legs of the angle.	ray / angle / vertex / ray
The size of the angle is measured in degrees. A right angle always measures 90 degrees. This is because a straight line measures 180 degrees. So if you divide the line in the middle by creating a vertical line, you will get two 90-degree angles.	180° 90°
An obtuse angle is greater than 90 degrees, but less than 180 degrees.	110°
An acute angle is less than 90 degrees.	70°

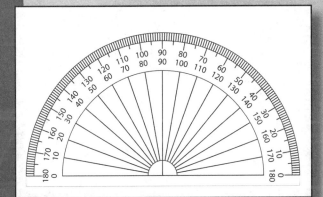

Tools Used to Measure Angles

There are many tools that can be used to measure angles. The most common tool is the **protractor**.

This protractor is a semi-circle. It is marked from 0–180 degrees. The protractor is the tool used most often to find the degrees of an angle.

57

Using a Protractor

We most often use a protractor to measure angles. Using a protractor is quite simple as long as you follow the correct steps.

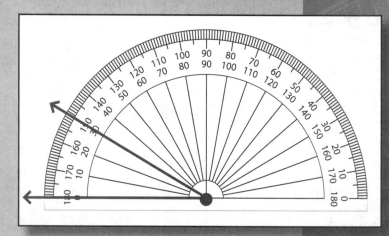

1. Always place the endpoint of the rays, or vertex of the angle on the center of the protractor.

2. Line the protractor up with one of the rays.

3. Look at the number of degrees on the protractor that lines up with the second ray. This will tell you the measure of the angle.

What is the measurement of the angle pictured above?

Remember, obtuse angles are larger than 90 degrees and acute angles are smaller than 90 degrees. So, you should use the row of numbers on the protractor that would fit that type of angle.

Measuring Angles in Our Daily Lives

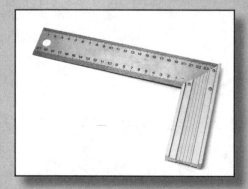

There are many people who measure angles in their jobs. Architects and engineers must measure angles. And construction workers also measure angles. If these angles were measured incorrectly, then we would have crooked houses and buildings! Many construction workers use set squares and miter saws to measure and cut wood at angles. If you look at your own home, you will see the many angles that were measured when it was being built!

You Try It

Identify each angle below as either an obtuse, acute, or right angle. Then use a protractor to get the exact measurements of the three angles.

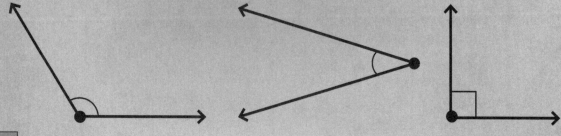

#50754—Leveled Texts for Mathematics: Measurement

Measuring Angles

How many angles are in this shape?
Are you certain that you counted them all?

What Are Angles?

An angle is what is formed when two rays meet. When the rays meet, they form a vertex and the two rays themselves form the legs of the angle.	ray / angle / vertex / ray
The size of the angle is measured in degrees. A right angle always measures 90 degrees. This makes sense because a straight line measures 180 degrees, so if you divide the line in the middle by creating a vertical perpendicular line, you will get two 90-degree angles.	180° 90°
An obtuse angle is greater than 90 degrees, but less than 180 degrees.	110°
An acute angle is less than 90 degrees.	70°

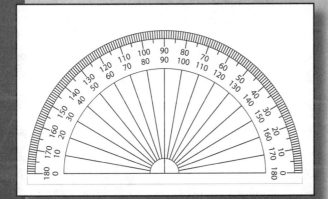

Tools Used to Measure Angles

The **protractor** is the most common of the many tools that can be used to measure angles.

This protractor is a semi-circle that is marked from 0–180 degrees. The protractor is the tool used most often to find the degrees of an angle.

#50754—Leveled Texts for Mathematics: Measurement

Using a Protractor

Using a protractor to measure angles is quite simple as long as you follow the correct steps.

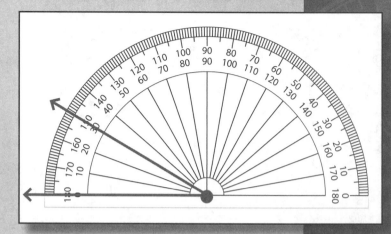

1. Always place the endpoint of the rays, or vertex of the angle on the center of the protractor.

2. Line the protractor up with one of the rays.

3. Look at the number of degrees on the protractor that lines up with the second ray to tell you the measure of the angle.

What is the measurement of the angle pictured above?

Remember, obtuse angles are larger than 90 degrees and acute angles are smaller than 90 degrees. When using a protractor confirm that you are using the row of numbers on the protractor that would fit that type of angle.

Measuring Angles in Our Daily Lives

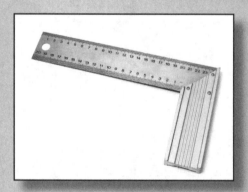

There are many people who measure angles in their jobs. Architects, engineers and construction workers must all measure angles. If these angles were measured incorrectly, then we would have crooked houses and buildings! Many construction workers use set squares and miter saws to measure and cut wood at angles. If you examine your own home, you will see the many angles that were measured when it was being built!

You Try It

Identify each angle below as either an obtuse, acute, or right angle. Then use a protractor to get the exact measurements of the three angles.

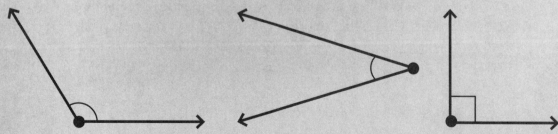

Estimating Measurements

Think of the African elephant. It is the largest living land animal. It can weigh up to 7,480 kilograms. That is 16,500 pounds. That is big! It stands from 3 to 4 meters tall at the shoulder. That is 10 to 13 feet. That is tall!

How Do We Estimate Measurements?

Estimate means to make a guess. But, it is a guess based on information. What if you don't have the right tools to measure? Or what if you do not need an exact answer? You can estimate. There are different ways to do this. They let you come close to the exact answer.

The Object	What It Measures
two fingers	about an inch in width
sheet of paper	about a foot in length
doorway	about a yard in width
nickel	about a gram in mass
loaf of bread	about a pound in weight

A loaf of bread weighs about a pound. About how many loaves of bread does your book weigh? First pick up the bread. Then pick up the book. Compare their weights. This lets you estimate. It tells about how many loaves of bread the book weighs.

You can do the same for length. It is easy to find the length in inches. You can use your fingers as a tool. Use them to measure. Try to measure your book. Then, measure the length using a ruler. This will help you see how close the estimate was.

Estimating lets us compare things. This lets us get a closer idea. You compared the weight of the bread and your book. This gave you a better idea about how much a book weighs. You used your fingers to measure your book. This lets you tell its length.

#50754—Leveled Texts for Mathematics: Measurement

Problems with Estimating Measurements

Estimating is a great way to find the measurement of things. This is true when you don't have the right tools. It is true when you don't need to be exact. But, it can be hard. It can cause problems. Use your fingers to see how many inches your desk is. Now have an adult do the same. Were the estimates the same?

Estimating Measurements in Our Daily Lives

There are times when we have to estimate. Think of people planning a trip. They may estimate the distance of the trip. They may need to stop for gas. They may need to stop for food. They can estimate how many miles they can go before stopping.

The map shows a city. Look at the scale on the map. Use the scale to estimate. Find the Ferris wheel downtown. Now find the Ferris wheel across the river. How many miles is it between them?

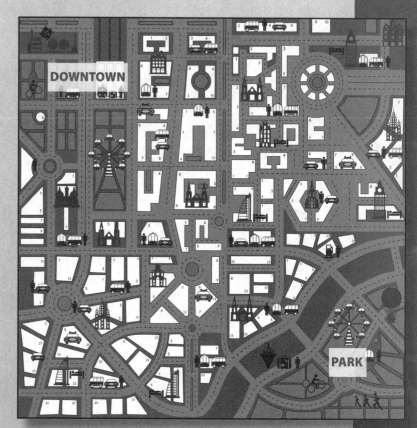

Scale

⊢—⊣ = 1 mi.

You Try It

Estimate the length of the things below. Then use a ruler to find the exact lengths.

- a pencil

- the seat of your chair

- the height of a bookcase in your classroom

Estimate the weight of the items below. Then use a scale to find the exact weight.

- a piece of chalk

- a globe

- a friend

#50754—Leveled Texts for Mathematics: Measurement

Estimating Measurements

Think of the African elephant. Did you know it is the largest living land animal? It can weigh up to 7,480 kilograms (16,500 pounds). It stands between 3 and 4 meters (10 and 13 feet) tall at the shoulder.

How Do We Estimate Measurements?

Estimate means to make an informed guess. What if you don't have the right tools to measure? Or what if you do not need an exact measurement? You can estimate. There are different ways to do this. They let you come close to the exact answer.

The Object	What It Measures
two fingers	about an inch in width
sheet of paper	about a foot in length
doorway	about a yard in width
nickel	about a gram in mass
loaf of bread	about a pound in weight

A loaf of bread weighs about a pound. About how many loaves of bread does your book weigh? First pick up the bread. Then pick up the book. You can compare the weights of both. You estimate about how many loaves of bread the book weighs.

You can do the same for length. It is easy to find the length in inches of your book. You can use your fingers as a tool. Use them to measure. Then measure the length using a ruler. This will help you see how close the estimate was.

Estimating lets us compare things. This gets an idea of their measurements. You compare the weight of the bread to the weight of the book. This gave you an idea about how much a book weighs. And, by using your fingers to measure your book, you could tell its length.

(63)

Problems with Estimating Measurements

Estimating is a great way to find the measurement of things when you don't have the right tools. It is also good when you don't need an exact measurement. But estimating can be hard. Use your fingers to see how many inches your desk is. Now have an adult do the same. Were the estimations the same?

Estimating Measurements in Our Daily Lives

There are times when we have to estimate. When people plan a trip, they estimate the distance of the trip. They may estimate how many miles they can travel before stopping for gas or food.

The map shows a city. Look at the scale on the map. Use the scale to estimate. How many miles is it from the Ferris wheel downtown to the Ferris wheel across the river?

You Try It

Estimate the length of the things below. Then use a ruler to find the exact lengths.

- a pencil

- the seat of your chair

- the height of a bookcase in your classroom

Estimate the weight of the items below. Then use a scale to find the exact weight.

- a piece of chalk

- a globe

- a friend

Scale

⊢—⊣ = 1 mi.

#50754—*Leveled Texts for Mathematics: Measurement*

Estimating Measurements

Did you know that the African elephant is the largest living land animal? It can weigh up to 7,480 kilograms (16,500 pounds). It stands between 3 and 4 meters (10 and 13 feet) tall at the shoulder.

How Do We Estimate Measurements?

Estimate means to make an informed guess. When you don't have the proper tools to measure or do not need an exact measurement, you must estimate. There are different ways to estimate so that you can come close to the exact answer.

The Object	What It Measures
two fingers	about an inch in width
sheet of paper	about a foot in length
doorway	about a yard in width
nickel	about a gram in mass
loaf of bread	about a pound in weight

You know that a loaf of bread weighs about a pound. About how many loaves of bread does your textbook weigh? First pick up the loaf of bread. Then pick up the textbook. By comparing the weights of both items, you can estimate about how many loaves of bread the textbook weighs.

You can do the same for length. It is easy to find the length in inches of your book by using your fingers as a measurement tool. Then you can measure the length using a ruler. This will help you see how accurate the estimate was.

Estimating lets us compare objects to get an idea of their measurements. By comparing the weight of the loaf of bread to the weight of your textbook, you were able to get a better idea about how much a textbook weighs. And by using your fingers to measure your textbook, you could determine its length.

65

Problems with Estimating Measurements

Estimating is a great way to find the measurement of items when you don't have the proper tools available or you don't need an exact measurement. But estimating can be a bit difficult. Use your fingers to estimate the width of your desk in inches. Now have an adult do the same. Were the estimations the same?

Estimating Measurements in Our Daily Lives

There are times when we have to estimate measurements. When people plan a trip, they estimate the distance of the trip. They also may estimate how many miles they can travel before stopping for gas or food.

The map shows a city. Look at the scale on the map. Use the scale to estimate how many miles it is from the Ferris wheel downtown to the Ferris wheel across the river.

You Try It

Estimate the length of the things below. Then use a ruler to find the exact lengths.

- a pencil

- the seat of your chair

- the height of a bookcase in your classroom

Estimate the weight of the items below. Then use a scale to find the exact weight.

- a piece of chalk

- a globe

- a friend

Scale

⊢—⊣ = 1 mi.

#50754—Leveled Texts for Mathematics: Measurement

© Shell Education

Estimating Measurements

Did you know that the African elephant is the largest living land animal? It can weigh up to 7,480 kilograms (16,500 pounds) and stands between 3 and 4 meters (10 and 13 feet) tall at the shoulder.

How Do We Estimate Measurements?

Estimate means to make an informed guess. When you don't have the proper tools to measure or do not need an exact measurement, estimation is a good strategy. There are a variety of ways to estimate so that you can approximate the exact answer.

The Object	What It Measures
two fingers	about an inch in width
sheet of paper	about a foot in length
doorway	about a yard in width
nickel	about a gram in mass
loaf of bread	about a pound in weight

A loaf of bread weighs about a pound; use this to estimate the weight of other objects. About how many loaves of bread does your textbook weigh? First pick up the loaf of bread, then pick up the textbook. By comparing the weights of both items, you can estimate how many loaves of bread would make up the weight of the textbook.

You can do the same for length. It is easy to find the length in inches of your book by using your fingers as a measurement tool. Then if you measure the length using a ruler, it will help you see how accurate the estimate was.

Estimating lets us compare objects to get a better idea of their measurements. By comparing the weight of the loaf of bread to the weight of your textbook, you were able to get a better idea about how much a textbook weighs. And by using your fingers to measure your textbook, you could determine its length.

67

Problems with Estimating Measurements

Estimating is a great way to find the measurement of items when you don't have the proper tools available or you don't need an exact measurement. But estimating has some inherent difficulties. Use your fingers to estimate how many inches wide your desk is, and then have an adult do the same. Were the estimations identical?

Estimating Measurements in Our Daily Lives

There are times when we have to estimate measurements. When people plan a trip, they estimate the distance of the trip. They also may estimate how many miles they can travel before stopping for gas or food.

The map shows a city. Look at the scale on the map and use it to estimate how many miles it is from the Ferris wheel downtown to the Ferris wheel across the river.

You Try It

Estimate the length of the things below. Then use a ruler to find the exact lengths.

- a pencil

- the seat of your chair

- the height of a bookcase in your classroom

Scale

⊢—⊣ = 1 mi.

Estimate the weight of the items below. Then use a scale to find the exact weight.

- a piece of chalk

- a globe

- a friend

Converting Length

A football field is 100 yards. How many feet is that? A team starts at one end. They run to the other end. How far do they go? They go 300 feet.

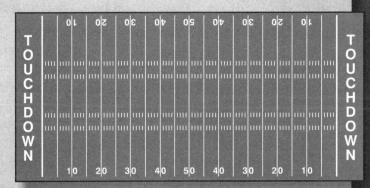

Rules for Converting Length in the English Units System

Measurement	Conversion
1 foot	12 inches
1 yard	36 inches or 3 feet
1 mile	5,280 feet or 1,760 yards

English Units Conversion Rule One: Divide to convert a small unit to a bigger one.

Jon walked 6,600 feet. How many miles is this? You can find out. Think of the rule. You need to divide. There are 5,280 feet in a mile. Divide 6,600 by 5,280. This gives you 1.25 miles. That is that same as $1\frac{1}{4}$ miles.

$$6,600 \text{ feet} \div \frac{1 \text{ mile}}{5,280 \text{ feet}} = 1.25 \text{ miles}$$

English Units Conversion Rule Two: Multiply to convert a big unit to a smaller one.

A field is 100 yards. How do you find how many feet that is? There are 3 feet in one yard. Think of the rule. You multiply.

$$100 \text{ yards} \times \frac{3 \text{ feet}}{1 \text{ yard}} = 300 \text{ feet}$$

Converting Length in the Metric System

Converting metric units can be easy. Metric uses groups of 10. How do you convert big units? You divide by groups of 10. How do you convert to small units? You multiply by multiples of 10. Tens make math easy!

There is a base unit in the metric system. What is the base unit? For length it is the meter. Big lengths are measured in kilometers. Small ones use millimeters.

 #50754—Leveled Texts for Mathematics: Measurement

Rules for Converting Length in the Metric System

Metric Unit	Unit Multiple
kilometers	1,000 (one thousand)
hectometers	100 (one hundred)
decameters	10 (ten)
Base Unit: meter	1
decimeters	0.1 (one-tenth)
centimeters	0.01 (one-hundredth)
millimeters	0.001 (one-thousandth)

Metric Conversion Rule One: Divide to convert from a small unit to a bigger one.

Two friends live 503,290 millimeters apart. How close is that? Millimeters are small. You can convert that to meters. Think of the rule. Divide by 1,000. You can see this on the chart. Millimeters and meters are 3 units apart. This is 503.290 meters. That is close to 500 meters.

$$503{,}290 \text{ millimeters} \div \frac{1 \text{ meter}}{1{,}000 \text{ millimeters}} = 503.29 \text{ meters}$$

Metric Conversion Rule Two: Multiply to convert a big unit to a smaller one.

Lina ran 1.5 kilometers. You can convert that to meters. Think of the rule. Multiply by 1,000. That is 1,500 meters.

$$1.5 \text{ kilometers} \times \frac{1{,}000 \text{ meters}}{1 \text{ kilometer}} = 1{,}500 \text{ meters}$$

Converting Length in Our Daily Lives

Many people convert lengths. They do this for their jobs. Designers draw clothes. Then they make them. They convert cloth lengths from yards to inches. Or they change them back. That is how they know how much cloth they need. How do painters know how much paint to buy? They convert the measurements of a room. How far must a runner run? They convert yards to miles. That is how they know how far to run.

You Try It

Convert these units:

7 feet = _____ inches

1,365 millimeters = _____ meters

#50754—Leveled Texts for Mathematics: Measurement

Converting Length

A football field is 100 yards. How many feet is that? How far does a team go to get from one end to the other end? They must go 300 feet.

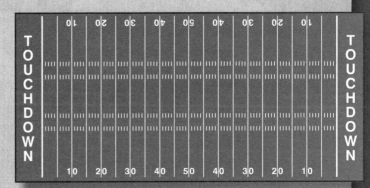

Rules for Converting Length in the English Units System

Measurement	Conversion
1 foot	12 inches
1 yard	36 inches or 3 feet
1 mile	5,280 feet or 1,760 yards

English Units Conversion Rule One: To convert a small unit to a bigger one, divide.

Jon walked 6,600 feet. You can find out the miles he went. Do this by dividing. This is because there are 5,280 feet in a mile. This gives you 1.25 miles. That is the same as 1 and $\frac{1}{4}$ miles.

$$6{,}600 \text{ feet} \div \frac{1 \text{ mile}}{5{,}280 \text{ feet}} = 1.25 \text{ miles}$$

English Units Conversion Rule Two: To convert a big unit to a smaller one, multiply.

A football field is 100 yards. There are 3 feet in one yard. How do you find how many feet are in the field? You multiply.

$$100 \text{ yards} \times \frac{3 \text{ feet}}{1 \text{ yard}} = 300 \text{ feet}$$

Converting Length in the Metric System

Converting units in the metric system can be easy. Metric is based on groups of 10. How do you convert into bigger units? You divide by multiples of 10. How do you convert into smaller units? You multiply by multiples of 10. Tens make for easy math!

The base unit is what you are using to measure. So the base unit of length in the metric system is the meter. This means that the largest lengths are measured in kilometers. The smallest lengths are measured in millimeters.

Rules for Converting Length in the Metric System

Metric Unit	Unit Multiple
kilometers	1,000 (one thousand)
hectometers	100 (one hundred)
decameters	10 (ten)
Base Unit: meter	1
decimeters	0.1 (one-tenth)
centimeters	0.01 (one-hundredth)
millimeters	0.001 (one-thousandth)

Metric Conversion Rule One: To convert from a small unit to a bigger one, divide.

Two friends live 503,290 millimeters apart. You can convert that to meters. Divide by 1,000. You can see this on the chart. Millimeters and meters are separated by 3 units. This would give you 503.290 meters. That is close to 500 meters.

$$503{,}290 \text{ millimeters} \div \frac{1 \text{ meter}}{1{,}000 \text{ millimeters}} = 503.29 \text{ meters}$$

Metric Conversion Rule Two: To convert a larger unit to a smaller unit, you should multiply.

Lin ran 1.5 kilometers. You can convert that to meters. Multiply by 1,000. That is 1,500 meters.

$$1.5 \text{ kilometers} \times \frac{1{,}000 \text{ meters}}{1 \text{ kilometer}} = 1{,}500 \text{ meters}$$

Converting Length in Our Daily Lives

Many people convert lengths in the real world. Designers convert fabric from yards to inches. Painters convert the measurements of a room. This is how they know how much paint to buy. Track runners convert yards to miles. That is how they know how far they have to run.

You Try It

Convert these units:

7 feet = _____ inches

1,365 millimeters = _____ meters

Converting Length

A football field is 100 yards. That means a team must travel 300 feet in order to get from one end zone to the other.

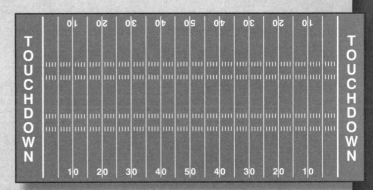

Rules for Converting Length in the English Units System

Measurement	Conversion
1 foot	12 inches
1 yard	36 inches or 3 feet
1 mile	5,280 feet or 1,760 yards

English Units Conversion Rule One: To convert a smaller unit to a larger unit, you should divide.

Jon walked 6,600 feet. You can find out the miles he walked by dividing. This is because there are 5,280 feet in a mile. This would give you 1.25 or $1\frac{1}{4}$ miles.

$$6,600 \text{ feet} \div \frac{1 \text{ mile}}{5,280 \text{ feet}} = 1.25 \text{ miles}$$

English Units Conversion Rule Two: To convert a larger unit to a smaller unit, you should multiply.

A football field is 100 yards. There are 3 feet in one yard. You should multiply to get the number of feet in a football field.

$$100 \text{ yards} \times \frac{3 \text{ feet}}{1 \text{ yard}} = 300 \text{ feet}$$

Converting Length in the Metric System

Converting units in the metric system can be easy. This is because the metric system is based on multiples of 10. You have to divide or multiply by multiples of 10 to convert units into either larger or smaller units.

The base unit is the unit you are using to measure. So the base unit of length in the metric system is meters. This means that larger lengths are measured in kilometers and smaller lengths are measured in millimeters.

73

Rules for Converting Length in the Metric System

Metric Unit	Unit Multiple
kilometers	1,000 (one thousand)
hectometers	100 (one hundred)
decameters	10 (ten)
Base Unit: meter	1
decimeters	0.1 (one-tenth)
centimeters	0.01 (one-hundredth)
millimeters	0.001 (one-thousandth)

Metric Conversion Rule One: To convert from a smaller unit to a larger unit, you should divide.

Two friends live 503,290 millimeters apart. You can convert that to meters by dividing by 1,000. You divide by 1,000 because millimeters and meters are separated by 3 units in the chart above. This would give you 503.290 meters, or about 500 meters.

$$503{,}290 \text{ millimeters} \div \frac{1 \text{ meter}}{1{,}000 \text{ millimeters}} = 503.29 \text{ meters}$$

Metric Conversion Rule Two: To convert a larger unit to a smaller unit, you should multiply.

Lin ran 1.5 kilometers. You can convert that to meters by multiplying by 1,000. This would give you 1,500 meters.

$$1.5 \text{ kilometers} \times \frac{1{,}000 \text{ meters}}{1 \text{ kilometer}} = 1{,}500 \text{ meters}$$

Converting Length in Our Daily Lives

There are many times when you convert lengths in the real world. Designers often have to convert fabric from yards to inches. Painters often convert the measurements of a room. They do this to decide how much paint to buy. Track runners convert yards to miles. They do this to see how far they have to run in their workouts.

You Try It

Convert the following units of measurement:

7 feet = _____ inches

1,365 millimeters = _____ meters

#50754—*Leveled Texts for Mathematics: Measurement*

Converting Length

A football field is 100 yards, which means a team must travel 300 feet in order to get from one end zone to the other.

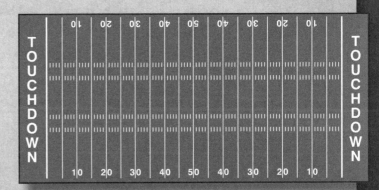

Rules for Converting Length in the English Units System

Measurement	Conversion
1 foot	12 inches
1 yard	36 inches or 3 feet
1 mile	5,280 feet or 1,760 yards

English Units Conversion Rule One: To convert a smaller unit to a larger unit, you should divide.

Jon walked 6,600 feet. You can find out the miles he walked by dividing because there are 5,280 feet in a mile. This would give you 1.25 or $1\frac{1}{4}$ miles.

$$6,600 \text{ feet} \div \frac{1 \text{ mile}}{5,280 \text{ feet}} = 1.25 \text{ miles}$$

English Units Conversion Rule Two: To convert a larger unit to a smaller unit, you should multiply.

A football field is 100 yards. There are 3 feet in one yard. You should multiply to get the number of feet in a football field.

$$100 \text{ yards} \times \frac{3 \text{ feet}}{1 \text{ yard}} = 300 \text{ feet}$$

Converting Length in the Metric System

Converting units in the metric system can be easy because the metric system is based on multiples of 10. You have to divide or multiply by multiples of 10 to convert units to larger or smaller units.

The base unit is the unit you are using to measure. There is a base unit of length in the metric system. It is the meter. This means that larger lengths are measured in kilometers and smaller lengths are measured in millimeters.

75

Rules for Converting Length in the Metric System

Metric Unit	Unit Multiple
kilometers	1,000 (one thousand)
hectometers	100 (one hundred)
decameters	10 (ten)
Base Unit: meter	1
decimeters	0.1 (one-tenth)
centimeters	0.01 (one-hundredth)
millimeters	0.001 (one-thousandth)

Metric Conversion Rule One: To convert from a smaller unit to a larger unit, you should divide.

Two friends live 503,290 millimeters apart. You can convert that to meters by dividing by 1,000. You divide by 1,000 because millimeters and meters are separated by 3 units in the chart above. This would give you 503.290 meters, or about 500 meters.

$$503{,}290 \text{ millimeters} \div \frac{1 \text{ meter}}{1{,}000 \text{ millimeters}} = 503.29 \text{ meters}$$

Metric Conversion Rule Two: To convert a larger unit to a smaller unit, you should multiply.

Lin ran 1.5 kilometers. You can convert that to meters by multiplying by 1,000. This would give you 1,500 meters.

$$1.5 \text{ kilometers} \times \frac{1{,}000 \text{ meters}}{1 \text{ kilometer}} = 1{,}500 \text{ meters}$$

Converting Length in Our Daily Lives

There are many times when you convert lengths in the real world. Designers often have to convert fabric from yards to inches, painters often convert the measurements of a room to decide how much paint to buy, and track runners convert yards to miles to figure out how far they have to run in their workouts.

You Try It

Convert the following units of measurement:

7 feet = _____ inches

1,365 millimeters = _____ meters

Converting Weight

You want to cook. You need things. You make a shopping list. You put 1 pound of flour. You put 2 pounds of rice. And you put 32 ounces of sugar. Then you go to the store. The bags say different amounts! There are 24 ounces of flour. There are 18 ounces of rice. And there are 2 pounds of sugar. What if you buy those bags? Will you have enough? Will you need to buy 2 bags for some things? How can you tell?

Do you know how to convert weight? This would help you. There are rules you can follow. They let you convert from one unit to another.

Converting Weights in English Units

Measurement	Conversion
1 pound	16 ounces or .0005 tons
1 ton	2,000 pounds or 32,000 ounces

English Units Conversion Rule One: To convert a small unit to a bigger one, divide.

You need 32 ounces of sugar. There are 16 ounces in a pound. You should divide 32 by 16. This shows that 32 ounces is 2 pounds. The bag is the right size.

$$32 \text{ ounces} \div \frac{1 \text{ pound}}{16 \text{ ounces}} = 2 \text{ pounds}$$

English Units Conversion Rule Two: To convert a big unit to a smaller one, multiply.

One pound of flour is 16 ounces. Two pounds of rice is 32 ounces. How do you know this? You multiply the 2 pounds you need by 16. What about your list? One bag of flour would be enough. Would one bag of rice be enough?

$$2 \text{ pounds} \times \frac{16 \text{ ounces}}{1 \text{ pound}} = 32 \text{ ounces}$$

Converting Weights in the Metric System

Converting units in metric can be easy. The metric system is based on groups of 10. You have to multiply by 10s. Or you need to divide by 10s. This will make units either larger or smaller.

What is a base unit? It is what you are using to measure. The base unit of mass is the gram. This is for the metric system. Big things are measured in kilograms. For small things, use milligrams.

Converting Weights in the Metric System (cont.)

Metric Unit	Unit Multiple
kilograms	1,000 (one thousand)
hectograms	100 (one hundred)
decagrams	10 (ten)
Base Unit: gram	1
decigrams	0.1 (one-tenth)
centigrams	0.01 (one-hundredth)
milligrams	0.001 (one-thousandth)

Metric Conversion Rule One: To convert from a small unit to a bigger one, divide.

Luke is 95,002,200 milligrams. You can convert that to kilograms. Divide by 1,000,000. Why? Do this because milligrams and kilograms are 6 units apart. You get 95.002200 kilograms. That is close to 95 kilograms. That is how much Luke weighs.

$$95,002,200 \text{ milligrams} \div \frac{1 \text{ kilogram}}{1,000,000 \text{ milligrams}} = 95 \text{ kilograms}$$

Metric Conversion Rule Two: To convert from a big unit to a smaller one, multiply.

You have a bag. It is 5.3 kilograms. You want to find out how many grams that is. You multiply by 1,000. This would give you 5,300 grams.

$$5.3 \text{ kilograms} \times \frac{1,000 \text{ gram}}{1 \text{ kilogram}} = 5,300 \text{ grams}$$

Converting Weight in Our Daily Lives

You are in your car. There is a bridge. You see a sign. You read it. "Take care! Top Weight is 4,000 pounds." Your car is less than two tons. But is that too heavy? How many pounds is that? You need to convert from tons to pounds. You want to make sure that the car is not too heavy. Think of when you cook. You may need to convert ounces to pounds. People convert weight all the time. They do it every day!

You Try It

Convert these units:

1.5 tons = _____ pounds

2,465 milligrams = _____ grams

Converting Weight

You make a shopping list. You put 1 pound of flour. You put 2 pounds of rice. And you put 32 ounces of sugar. When you get to the store you find that the bags say different amounts. There are 24 ounces of flour. There are 18 ounces of rice. And there are 2 pounds of sugar. If you buy those bags, will you have enough of each thing? Will you need to buy 2 bags for some of them?

Knowing how to convert weight would help you. There are different rules you can follow. They will help you to convert from one unit to another.

Converting Weights in English Units

Measurement	Conversion
1 pound	16 ounces or .0005 tons
1 ton	2,000 pounds or 32,000 ounces

English Units Conversion Rule One: To convert a small unit to a bigger one, divide.

You need 32 ounces of sugar. There are 16 ounces in a pound. So, you should divide 32 by 16. This shows that 32 ounces equals 2 pounds.

$$32 \text{ ounces} \div \frac{1 \text{ pound}}{16 \text{ ounces}} = 2 \text{ pounds}$$

English Units Conversion Rule Two: To convert a big unit to a smaller one, multiply.

One pound of flour is 16 ounces. Two pounds of rice would be 32 ounces. You multiply the 2 pounds you need by 16. In the problem above, one bag of flour would be enough. Would one bag of rice be enough?

$$2 \text{ pounds} \times \frac{16 \text{ ounces}}{1 \text{ pound}} = 32 \text{ ounces}$$

Converting Weight in the Metric System

Converting units in metric can be easy. This is because the metric system is based on groups of 10. You have to multiply by 10s. Or you need to divide by 10s. This will make units either larger or smaller.

The base unit is what you are using to measure. The base unit of mass in the metric system is the gram. Big weights are measured in kilograms. Small weights use milligrams.

#50754—Leveled Texts for Mathematics: Measurement

Converting Weights in the Metric System *(cont.)*

Metric Unit	Unit Multiple
kilograms	1,000 (one thousand)
hectograms	100 (one hundred)
decagrams	10 (ten)
Base Unit: gram	1
decigrams	0.1 (one-tenth)
centigrams	0.01 (one-hundredth)
milligrams	0.001 (one-thousandth)

Metric Conversion Rule One: To convert from a small unit to a bigger one, divide.

Luke weighs 95,002,200 milligrams. You can convert that to kilograms. Divide by 1,000,000. This is because milligrams and kilograms are 6 units apart. This gives you 95.002200 kilograms. That is close to 95 kilograms.

$$95{,}002{,}200 \text{ milligrams} \div \frac{1 \text{ kilogram}}{1{,}000{,}000 \text{ milligrams}} = 95 \text{ kilograms}$$

Metric Conversion Rule Two: To convert a big unit to a smaller unit, multiply.

You are carrying a bag. It weighs 5.3 kilograms. You want to find out how many grams that is. You multiply by 1,000. This would give you 5,300 grams.

$$5.3 \text{ kilograms} \times \frac{1{,}000 \text{ gram}}{1 \text{ kilogram}} = 5{,}300 \text{ grams}$$

Converting Weight in Our Daily Lives

You cross over a bridge. You see a sign. It reads, "Caution: Maximum Weight 4,000 pounds." You know that your car weighs less than two tons. You need to convert from tons to pounds. You want to make sure that the car is not too heavy. When cooking, you may need to convert ounces to pounds. There are many times when people convert weight every day!

You Try It

Convert the following units of measurement:

1.5 tons = _____ pounds

2,465 milligrams = _____ grams

Converting Weight

On your grocery list you have 1 pound of flour and 2 pounds of rice. You have 32 ounces of sugar. When you get to the store you find that the packages say 24 ounces of flour, 18 ounces of rice, and 2 pounds of sugar. If you buy those packages, will you have enough of each ingredient? Or will you need to buy extra packages for some of the items?

Knowing how to convert weight would help you solve this problem. There are different rules you can follow to convert from one unit of weight to another.

Converting Weights in English Units

Measurement	Conversion
1 pound	16 ounces or .0005 tons
1 ton	2,000 pounds or 32,000 ounces

English Units Conversion Rule One: To convert a smaller unit to a larger unit, you should divide.

You need 32 ounces of sugar. You should divide 32 ounces by 16. This is because there are 16 ounces in a pound. This shows that 32 ounces equals 2 pounds.

$$32 \text{ ounces} \div \frac{1 \text{ pound}}{16 \text{ ounces}} = 2 \text{ pounds}$$

English Units Conversion Rule Two: To convert a larger unit to a smaller unit, you should multiply.

One pound of flour is 16 ounces. Two pounds of rice would be 32 ounces. You multiply the 2 pounds you need by 16. In the problem above, one package of flour would be enough. Would one package of rice be enough?

$$2 \text{ pounds} \times \frac{16 \text{ ounces}}{1 \text{ pound}} = 32 \text{ ounces}$$

Converting Weight in the Metric System

Converting units in the metric system can be easy. This is because the metric system is based on multiples of 10. You have to multiply or divide by multiples of 10 to make units either larger or smaller.

The base unit is the unit you are using to measure. The base unit of mass in the metric system is the gram. This means that the largest weights are measured in kilograms and the smallest weights in milligrams.

#50754—Leveled Texts for Mathematics: Measurement

Converting Weights in the Metric System (cont.)

Metric Unit	Unit Multiple
kilograms	1,000 (one thousand)
hectograms	100 (one hundred)
decagrams	10 (ten)
Base Unit: gram	1
decigrams	0.1 (one-tenth)
centigrams	0.01 (one-hundredth)
milligrams	0.001 (one-thousandth)

Metric Conversion Rule One: To convert from a smaller unit to a larger unit, you must divide.

Luke weighs 95,002,200 milligrams. You can convert that to kilograms by dividing by 1,000,000. You divide by 1,000,000. This is because milligrams and kilograms are separated by 6 units. This would give you 95.002200 kilograms. That is about 95 kilograms.

$$95{,}002{,}200 \text{ milligrams} \div \frac{1 \text{ kilogram}}{1{,}000{,}000 \text{ milligrams}} = 95 \text{ kilograms}$$

Metric Conversion Rule Two: To convert a larger unit to a smaller unit, you must multiply.

You are carrying a bag that weighs 5.3 kilograms. You want to find out how many grams that is. You multiply by 1,000. This would give you 5,300 grams.

$$5.3 \text{ kilograms} \times \frac{1{,}000 \text{ gram}}{1 \text{ kilogram}} = 5{,}300 \text{ grams}$$

Converting Weight in Our Daily Lives

You cross over a bridge. You notice that the sign reads, "Caution: Maximum Weight 4,000 pounds." You know that your car weighs less than two tons. So you would need to convert from tons to pounds to make sure that the car is not over the weight limit. When cooking, you may need to covert ounces to pounds. There are many times when people convert weight every day!

You Try It

Convert the units of measurement:

1.5 tons = _____ pounds

2,465 milligrams = _____ grams

Converting Weight

On your grocery list you have 1 pound of flour, 2 pounds of rice, and 32 ounces of sugar. When you get to the store you find that the packages say 24 ounces of flour, 18 ounces of rice, and 2 pounds of sugar. If you buy those packages, will you have enough of each ingredient, or will you need to buy 2 packages for some of the items?

Knowing how to convert weight would help you solve this problem. There are different rules you can follow to convert from one unit of weight to another.

Converting Weights in English Units

Measurement	Conversion
1 pound	16 ounces or .0005 tons
1 ton	2,000 pounds or 32,000 ounces

English Units Conversion Rule One: To convert a smaller unit to a larger unit, you should divide.

You need 32 ounces of sugar. You should divide 32 ounces by 16 because there are 16 ounces in a pound. This shows that 32 ounces equals 2 pounds.

$$32 \text{ ounces} \div \frac{1 \text{ pound}}{16 \text{ ounces}} = 2 \text{ pounds}$$

English Units Conversion Rule Two: To convert a larger unit to a smaller unit, you should multiply.

One pound of flour is 16 ounces, so two pounds of rice would be 32 ounces. You multiply the 2 pounds you need by 16. In the problem above, one package of flour would be enough. Would one package of rice be enough?

$$2 \text{ pounds} \times \frac{16 \text{ ounces}}{1 \text{ pound}} = 32 \text{ ounces}$$

Converting Weight in the Metric System

Converting units in the metric system can be easy because the metric system is based on multiples of 10. You have to multiply or divide by multiples of 10 to make units either larger or smaller.

The base unit is the unit you are using to measure. The base unit of mass in the metric system is the gram, which means that the largest weights are measured in kilograms and the smallest weights in milligrams.

Converting Weights in the Metric System *(cont.)*

Metric Unit	Unit Multiple
kilograms	1,000 (one thousand)
hectograms	100 (one hundred)
decagrams	10 (ten)
Base Unit: gram	1
decigrams	0.1 (one-tenth)
centigrams	0.01 (one-hundredth)
milligrams	0.001 (one-thousandth)

Metric Conversion Rule One: To convert from a smaller unit to a larger unit, you must divide.

Luke weighs 95,002,200 milligrams. You can convert that to kilograms by dividing by 1,000,000. You divide by 1,000,000 because milligrams and kilograms are separated by 6 units, giving you 95.002200 kilograms, or about 95 kilograms.

$$95{,}002{,}200 \text{ milligrams} \div \frac{1 \text{ kilogram}}{1{,}000{,}000 \text{ milligrams}} = 95 \text{ kilograms}$$

Metric Conversion Rule Two: To convert a larger unit to a smaller unit, you must multiply.

You are carrying a bag that weighs 5.3 kilograms and you want to find out how many grams that is. You multiply by 1,000, which would give you 5,300 grams.

$$5.3 \text{ kilograms} \times \frac{1{,}000 \text{ gram}}{1 \text{ kilogram}} = 5{,}300 \text{ grams}$$

Converting Weight in Our Daily Lives

You cross over a bridge. You notice that the sign reads, "Caution: Maximum Weight 4,000 pounds." You know that your car weighs less than two tons, so you would need to convert from tons to pounds to make sure that the car is not over the weight limit. When cooking, you may need to covert ounces to pounds. There are many times when we convert weight every day!

You Try It

Convert the units of measurement:

1.5 tons = _____ pounds

2,465 milligrams = _____ grams

84

Measuring the Perimeter of Regular Shapes

Have you walked around the outside of a playground? Did you count how many steps it took? As you were counting steps, you were measuring the perimeter, too!

Basic Facts

The **perimeter** of a shape is the border. It is the outside boundary. Perimeters can be measured on two-dimensional things. Finding the perimeter of a regular shape is easy. A regular shape has sides that are the same length. It has angle measures that are the same, too.

Finding the Perimeter of Any Regular Shape

Here are the steps to find the perimeter of a regular shape.

1. Measure the length of one side of the shape.  Each side = 4 cm	**2.** Add all of the side lengths together or multiply the side length by the number of sides.  4 cm + 4 cm + 4 cm = 12 cm or 3 x 4 cm = 12 cm

Here are some equations. Follow them to measure the perimeter of regular shapes (L = length of one side).

Perimeter of a regular triangle	$L + L + L$ or $3 \times L$
Perimeter of a square	$L + L + L + L$ or $4 \times L$
Perimeter of a regular pentagon	$L + L + L + L + L$ or $5 \times L$
Perimeter of a regular hexagon	$L + L + L + L + L + L$ or $6 \times L$

85

Using the Equations

The equations make it easy. They let you figure the perimeters. Look at these regular shapes. The equations are used. They help to figure the perimeter of the square. They help with the pentagon, too.

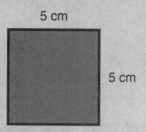

5 cm

5 cm

$$5 \text{ cm} + 5 \text{ cm} + 5 \text{ cm} + 5 \text{ cm} = 20 \text{ cm}$$

$$4 \times 5 \text{ cm} = 20 \text{ cm}$$

6 in.

$$6 \text{ in.} + 6 \text{ in.} + 6 \text{ in.} + 6 \text{ in.} + 6 \text{ in.} = 30 \text{ in.}$$

$$5 \times 6 \text{ in.} = 30 \text{ in.}$$

Measuring Perimeter in Our Daily Lives

Many people have pet dogs. Dogs need fenced-in places to play. They need to be safe. How can you figure out the length of the fence? You need to measure the perimeter. What if you want to redecorate a room? You need to know how big the room is. You need to know the perimeter. This is so you can get the best things to fit the room. There are many times when perimeter is used in the real world. Can you name one?

You Try It

Find the perimeter of an octagon that measures 8 inches on each side.

8 in.

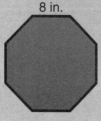

#50754—Leveled Texts for Mathematics: Measurement

Measuring the Perimeter of Regular Shapes

Have you walked around the outside of a playground and counted to see how many steps it would take? Did you know that as you were counting steps, you were measuring the perimeter, too?

Basic Facts

The **perimeter** of a shape is the border, or outside boundary. Perimeters can be measured on two-dimensional shapes. Finding the perimeter of a regular shape is easy to do. A regular shape has sides that are the same length. It has angle measures that are the same, too.

Finding the Perimeter of Any Regular Shape

Here are the steps to find the perimeter of a regular shape:

1. Measure the length of one side of the shape. Each side = 4 cm	**2.** Add all of the side lengths together or multiply the side length by the number of sides. 4 cm + 4 cm + 4 cm = 12 cm or 3 x 4 cm = 12 cm

Here are some equations. Follow them to measure the perimeter of regular shapes (L = length of one side).

Perimeter of a regular triangle	$L + L + L$ or $3 \times L$
Perimeter of a square	$L + L + L + L$ or $4 \times L$
Perimeter of a regular pentagon	$L + L + L + L + L$ or $5 \times L$
Perimeter of a regular hexagon	$L + L + L + L + L + L$ or $6 \times L$

Using the Equations

Using the equations makes it easy to figure the perimeter of regular shapes. Look at these regular shapes below. The equations are used. They help to figure the perimeter of the square and the pentagon.

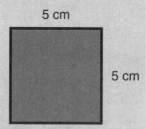

5 cm

5 cm

$$5 \text{ cm} + 5 \text{ cm} + 5 \text{ cm} + 5 \text{ cm} = 20 \text{ cm}$$

$$4 \times 5 \text{ cm} = 20 \text{ cm}$$

6 in.

$$6 \text{ in.} + 6 \text{ in.} + 6 \text{ in.} + 6 \text{ in.} + 6 \text{ in.} = 30 \text{ in.}$$

$$5 \times 6 \text{ in.} = 30 \text{ in.}$$

Measuring Perimeter in Our Daily Lives

Many people have pet dogs. Dogs need fenced-in places to play. How can you figure out the length of the fence? You need to measure the perimeter. What if you want to redecorate a room? You would need to know how big the room is. You would need to know the perimeter. This is so you can get the best furniture to fit the room. There are many times when perimeter is used in the real world. Can you name one?

You Try It

Find the perimeter of an octagon that measures 8 inches on each side.

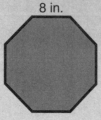

8 in.

#50754—*Leveled Texts for Mathematics: Measurement*

Measuring the Perimeter of Regular Shapes

Have you ever walked around the outside of a playground to see how many steps it would take? What you may not have realized was that as you were counting the steps, you were also measuring the perimeter.

Basic Facts

The **perimeter** of an object is the border or outside boundary. Finding the perimeter of a regular shape is easy because regular shapes have sides that are the same length and angle measures that are the same.

Finding the Perimeter of Any Regular Shape

Here are the steps to find the perimeter of a regular shape.

1. Measure the length of one side of the shape.  Each side = 4 cm	**2.** Add all of the side lengths together or multiply the side length by the number of sides. 4 cm + 4 cm + 4 cm = 12 cm or 3 x 4 cm = 12 cm

Here are some equations. Follow them to measure the perimeter of regular shapes (L = length of one side).

Perimeter of a regular triangle	$L + L + L$ or $3 \times L$
Perimeter of a square	$L + L + L + L$ or $4 \times L$
Perimeter of a regular pentagon	$L + L + L + L + L$ or $5 \times L$
Perimeter of a regular hexagon	$L + L + L + L + L + L$ or $6 \times L$

#50754—Leveled Texts for Mathematics: Measurement

Using the Equations

Using the equations to calculate the perimeter of regular shapes is easy to do. Look at the regular shapes below. The equations are used to calculate the perimeter of the square and the pentagon.

5 cm

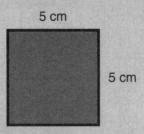

5 cm

$$5 \text{ cm} + 5 \text{ cm} + 5 \text{ cm} + 5 \text{ cm} = 20 \text{ cm}$$

$$4 \times 5 \text{ cm} = 20 \text{ cm}$$

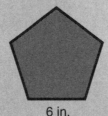

6 in.

$$6 \text{ in.} + 6 \text{ in.} + 6 \text{ in.} + 6 \text{ in.} + 6 \text{ in.} = 30 \text{ in.}$$

$$5 \times 6 \text{ in.} = 30 \text{ in.}$$

Measuring Perimeter in Our Daily Lives

Many people have pet dogs and it is important for dogs to have fenced-in places to play. The only way to figure out the length of the fence is by measuring the perimeter. The same is true to find the perimeter of a room you are redecorating since you would need to know how large the room is in order to get the best furniture that would fit. There are many examples of using perimeter in the real world. Can you name one?

You Try It

Find the perimeter of an octagon that measures 8 inches on each side.

8 in.

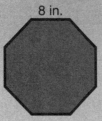

Measuring the Perimeter of Regular Shapes

Have you ever paced out the outside of a playground to figure out the number of steps? As you were counting the steps, you were also measuring the perimeter.

Basic Facts

The **perimeter** is the border or outside boundary. Perimeters can be measured on two-dimensional shapes. Finding the perimeter of a regular shape is simple because these have equivalent side lengths and angle measures.

Finding the Perimeter of Any Regular Shape

Here are the steps to find the perimeter of a regular shape.

1. Measure the length of one side of the shape.	2. Add all of the side lengths together or multiply the side length by the number of sides.
Each side = 4 cm	4 cm + 4 cm + 4 cm = 12 cm or 3 x 4 cm = 12 cm

Here are some equations. Follow them to measure the perimeter of regular shapes (L = length of one side).

Perimeter of a regular triangle	$L + L + L$ or $3 \times L$
Perimeter of a square	$L + L + L + L$ or $4 \times L$
Perimeter of a regular pentagon	$L + L + L + L + L$ or $5 \times L$
Perimeter of a regular hexagon	$L + L + L + L + L + L$ or $6 \times L$

91

Using the Equations

Using the equations to calculate the perimeter of regular shapes is very straightforward. Examine the regular shapes below and notice that the equations are used to calculate the perimeter of the square and the pentagon.

5 cm

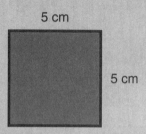

5 cm

$$5 \text{ cm} + 5 \text{ cm} + 5 \text{ cm} + 5 \text{ cm} = 20 \text{ cm}$$

$$4 \times 5 \text{ cm} = 20 \text{ cm}$$

6 in.

$$6 \text{ in.} + 6 \text{ in.} + 6 \text{ in.} + 6 \text{ in.} + 6 \text{ in.} = 30 \text{ in.}$$

$$5 \times 6 \text{ in.} = 30 \text{ in.}$$

Measuring Perimeter in Our Daily Lives

People with pet dogs know that it is imperative for these energetic animals to have fenced-in places to play. The only way to figure out the length of the fence needed is by measuring the perimeter of the play area. The same is true in finding the perimeter of a room you are redecorating. In order to determine which furniture pieces would best fit the room, it would be necessary to know the room's dimensions. There are many examples of using perimeter in the real world. Can you name one?

You Try It

Find the perimeter of an octagon that measures 8 inches on each side.

8 in.

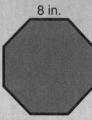

Measuring the Area of Regular Shapes

A room is 12 feet long. It is 12 feet wide. It is a square. You need to put carpet on the floor. You want enough to cover the whole floor. How much carpet do you need?

Basic Facts

Area is the number of square units needed to cover the space inside a shape. Area only measures two-dimensional shapes. Look at the square below. There are small squares inside of it. They show how much space the big square takes up. There are 9 smaller squares. The area of the big square is 9 square units. This can be written 9 u^2.

Triangles are trickier. How do you find their area? Look at the shaded area below. You have 3 whole squares. And you have partial squares. You have to add them together. The partial squares make 1.5 whole squares. There are 3 partial squares. If you add them all up, you get 4.5 whole squares. This means that the area is 4.5 square units or 4.5 u^2.

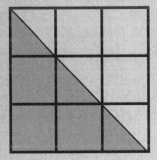

You can find the area of floors. Or pieces of paper. They have a length and a width. But not the area of books. They are 3-D. You find the surface area of 3-D things. Area is two-dimensional (2-D). So it is measured in square units. You must place a small 2 after the unit of measurement. This shows you are measuring area.

Using Equations

It can take a long time to count squares in a grid. So it can be easier to use an equation. Here are some equations to use. They can save you time.

Regular Shape	Equation	Diagram
square	area (A) = length × width	4 in. square, 4 in.
triangle	area (A) = base × height ÷ 2	6 m, 8 m triangle

To find the amount of carpet from the problem on the last page, you first have to find the area of the room. The room is square. The area of a square is found by multiplying length × width. So you would multiply 12 × 12. This gets you 144 feet^2.

Calculating Area of a Triangle

Look at the triangle below. The base is 4 meters. The height is 3 meters. Follow the triangle equation for area. Multiply 4 by 3. This is the base times the height. This gives you 12. Then, divide by 2. This gets 6 m^2.

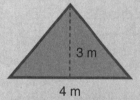

4 meters × 3 meters = 12 m^2

3 m

4 m

Measuring Area in Our Daily Lives

People who work at zoos must measure the area of the living spaces for the animals. This is important in order to make the best spaces for the animals. You have to know the area the animals would need to roam and be happy.

You Try It

Find the area of the shapes below:

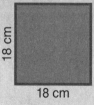

18 cm

18 cm

A = _____

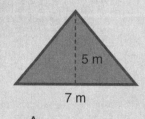

5 m

7 m

A = _____

Measuring the Area of Regular Shapes

A room is 12 feet long. It is 12 feet wide. You need enough carpet to cover the entire floor of the room. How much carpet do you need?

Basic Facts

Area is the number of square units needed to cover the space inside a figure. Area only measures two-dimensional surfaces. Look at the square below. There are small squares inside the square. They show how much space the big square takes up. There are 9 smaller squares. This means that the area of the big square is 9 square units. This can be written 9 u^2.

Triangles are a little trickier. How do you find their area? Look at the shaded area below. You have 3 whole squares. The partial squares make 1.5 whole squares. If you add them all up, you get 4.5 whole squares. This means that the area is 4.5 square units or 4.5 u^2.

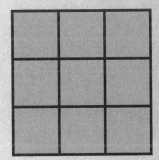

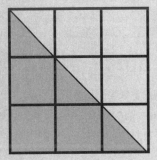

You can find the area of floors. Or of pieces of paper. They have a length and a width. But you can't find the area of books. They are three-dimensional. Instead you find the surface area of 3-D objects. Since area is two-dimensional, it is measured in square units. You must place a small 2 after the unit of measurement to show you are measuring area.

Using Equations

It can take a long time to count squares in a grid. So it can be easier to use an equation. Here are some equations to use.

Regular Shape	Equation	Diagram
square	area (A) = length × width	4 in. (sides) 4 in.
triangle	area (A) = base × height ÷ 2	6 m, 8 m

In order to find the amount of carpet needed from the problem on the previous page, you would first have to find the area of the room. The area of a square is found by multiplying length × width. So you would multiply 12 × 12 to get 144 feet^2.

Calculating Area of a Triangle

On the triangle below, the base is 4 meters and the height is 3 meters. So following the triangle equation for finding area, you would multiply 4 by 3, or the base times the height. This would give you 12. Then, divide by 2 to get 6 m^2.

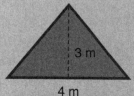
3 m, 4 m

4 meters × 3 meters = 12 m^2

Measuring Area in Our Daily Lives

People who work at zoos must measure the area of the living spaces for the animals. This is important in order to make the best spaces for the animals. You have to know the area the animals would need to roam and be happy.

You Try It

Find the area of the shapes below:

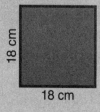
18 cm, 18 cm

A = _____

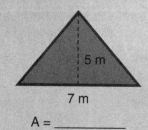
5 m, 7 m

A = _____

Measuring the Area of Regular Shapes

A room is 12 feet long by 12 feet wide. You have been assigned the task of getting enough carpet to cover the entire floor of the room. So how much carpet would you need?

Basic Facts

Area is the number of square units needed to cover the space inside a figure. Area only measures two-dimensional surfaces. Look at the square below. The small squares inside the square show how much space the square takes up. There are 9 smaller squares, which means that the area of the square is 9 square units or 9 u^2.

To find the area of a triangle is a little trickier. You have to add the number of partial squares to the number of complete squares. Look at the triangle below. There are 3 complete squares and partial squares make another 1.5 whole squares, which means that the area is 4.5 square units or 4.5 u^2.

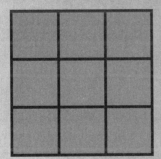

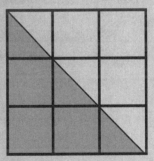

You can find the area of floors or pieces of paper. They have a length and a width. But, you can't find the area of books. They are three-dimensional. Instead, you would find the surface area of 3-D objects. Since area is two-dimensional, it is measured in square units. You must place a small 2 (like this: ^{2}) after the unit of measurement to show you are measuring area.

Using Equations

It can take a long time to count squares in a grid every time you want to find the area of a shape. So it is easier to use an equation instead. Here are some equations that can be used.

Regular Shape	Equation	Diagram
square	area (A) = length × width	4 in. / 4 in.
triangle	area (A) = base × height ÷ 2	6 m / 8 m

In order to figure out the amount of carpet needed from the problem on the previous page, you would first have to find the area of the room. The area of a square is found by multiplying length × width. So you would multiply 12 × 12 to get 144 feet².

Calculating Area of a Triangle

On the triangle below, the base is 4 meters and the height is 3 meters. So following the triangle equation for finding area, you would multiply 4 by 3, or the base times the height. This would give you 12. Then, divide by 2 to get 6 m².

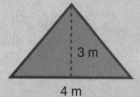

4 meters × 3 meters = 12 m²

Measuring Area in Our Daily Lives

People who work at zoos must measure the area of the living spaces for the animals. In order to design the best spaces for the animals, you would have to know the area the animals would need to roam comfortably.

You Try It

Find the area of the following shapes:

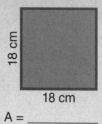

A = _____

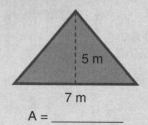

A = _____

Measuring the Area of Regular Shapes

You have been assigned the task of getting enough carpet to cover the entire floor of a room that is 12 feet long by 12 feet wide. So how do you determine how much carpet you would need?

Basic Facts

Area is the number of square units needed to cover the space inside a two-dimensional figure. Area only measures two-dimensional surfaces. Consider the square below. The small squares inside the larger square show how much space the square takes up. There are 9 smaller squares, which means that the area of the square is 9 square units or 9 u^2.

To find the area of a triangle is a little trickier. You have to add the number of partial squares to the number of complete squares. Consider the triangle below. There are 3 complete squares and partial squares make another 1.5 whole squares. The total area is 4.5 square units or 4.5 u^2.

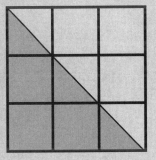

You can find the area of floors or pieces of paper because they have a length and a width. But, you can't find the area of books because they are three-dimensional. Instead, you would find the surface area of three-dimensional objects. Since area is two-dimensional, it is measured in square units, which means that you must place a small 2 (like this: ^{2}) after the unit of measurement. This indicates that you are measuring area.

Using Equations

It can take a long time to count squares in a grid every time you want to find the area of a shape, so it is often easier to use an equation, instead. Here are some equations that can be used.

Regular Shape	Equation	Diagram
square	area (A) = length × width	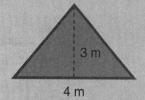 4 in. / 4 in.
triangle	area (A) = base × height ÷ 2	6 m / 8 m

In order to figure out the amount of carpet needed from the problem on the previous page, you would first have to find the area of the room. The area of a square is found by multiplying length × width, so you would multiply 12 × 12 to get 144 feet^2.

Calculating Area of a Triangle

On the triangle below, the base is 4 meters and the height is 3 meters. Following the triangle equation for finding area, you would multiply 4 by 3, or the base times the height, which would give you 12. Then divide by 2 to get 6 m^2.

4 meters × 3 meters = 12 m^2

3 m

4 m

Measuring Area in Our Daily Lives

People who work at zoos must measure the area of the living spaces for the animals. In order to design the best spaces for the animals, you would have to know the area the animals would need to roam comfortably.

You Try It

Find the area of the following shapes.

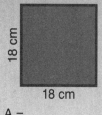

18 cm

18 cm

A = _____

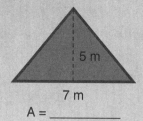

5 m

7 m

A = _____

100

Measuring the Perimeter of Irregular Shapes

Look at the shape below. How would you find its perimeter?

Basic Facts

Perimeter is a distance. It is the length around the outside of a shape. We measure perimeter in linear units. We use inches. We use centimeters. We use meters. We use feet. Linear units measure length.

Finding the perimeter of regular shapes is easy. You just measure one side of the shape. Then you multiply that length by the number of sides. How would you find the perimeter of this square? You multiply 4 sides × 4 meters. This gives you 16 meters.

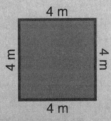

How would you find the perimeter of the pentagon? You multiply 5 sides × 3 inches. This gives you 15 inches.

Irregular shapes do not have sides that are all the same length. Some sides may be short. Some may be long. The angles can be different, too. Here are some irregular shapes.

#50754—Leveled Texts for Mathematics: Measurement

Steps for Measuring the Perimeter of Irregular Shapes

Finding the perimeter of irregular shapes takes more steps. The perimeter is the sum of the segments around the outside of the shape.

Follow these steps. They will let you find the perimeter of the shape at right:

1. Find the endpoints of each line segment.

2. Measure the length of each segment.

3. Add the lengths together. This will give you the perimeter of the irregular shape.

The line segments on the shape were measured. One side is 4 cm. One side is 2 cm. One side is 5 cm. One side is 9 cm. One side is 8 cm. The last side is 11 cm. Add those numbers together. You get a perimeter of 39 cm.

Measuring the Perimeter of Irregular Shapes in Our Daily Lives

Not all shapes are regular shapes. There are times when the perimeter of irregular shapes must be found. Pools can be irregular shapes. The perimeter must be measured before the ground is broken and concrete is poured.

When building a home, the floor plans can be irregular shapes. Contractors must be able to measure the perimeter of the home, too.

You Try It

Find the perimeter of the shapes below.

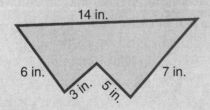

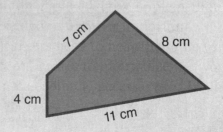

#50754—*Leveled Texts for Mathematics: Measurement*

© *Shell Education*

Measuring the Perimeter of Irregular Shapes

Look at the shape below. How would you find the perimeter of this shape?

Basic Facts

Perimeter is the distance around the outside of a shape. We measure perimeter in linear units. We use inches and centimeters. We use meters and feet. Linear units measure length.

Finding the perimeter of regular shapes is easy. You just measure one side of the shape. Then you multiply that length by the number of sides. To find the perimeter of this square, you multiply 4 sides × 4 meters. This gives you 16 meters.

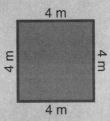

To find the perimeter of the pentagon below, you multiply 5 sides × 3 inches. This gives you 15 inches.

Irregular shapes do not have sides that are all the same length. Some sides may be short. Some may be long. Here are some examples of irregular shapes.

 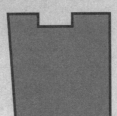 

#50754—Leveled Texts for Mathematics: Measurement

Steps for Measuring the Perimeter of Irregular Shapes

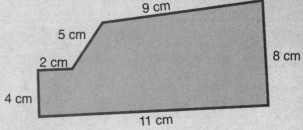

Finding the perimeter of irregular shapes takes more steps than for regular shapes. The perimeter equals the sum of the segments around the outside of the figure.

Follow these steps to find the perimeter of the irregular shape at right:

1. Find the endpoints of each line segment on the shape.

2. Measure the length of each segment.

3. Add the lengths together. This will give you the perimeter of the irregular shape.

On the shape above, the line segments measure 4 cm, 2 cm, 5 cm, 9 cm, 8 cm, and 11 cm. If you add those numbers together, you get a perimeter of 39 cm.

Measuring the Perimeter of Irregular Shapes in Our Daily Lives

Not all shapes in our world are regular shapes. There are times when the perimeter of irregular shapes must be found. Pools can be irregular shapes. The perimeter must be measured before the ground is broken and concrete is poured.

When building a home, the floor plans can be irregular shapes. Contractors must be able to measure the perimeter of the home, too.

You Try It

Find the perimeter of the following irregular shapes.

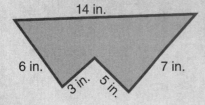

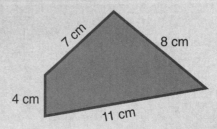

Measuring the Perimeter of Irregular Shapes

Look at the shape below. What would you need to do to find its perimeter?

Basic Facts

Perimeter is the distance around the outside of a figure. We measure perimeter in linear units such as inches, centimeters, meters, or feet. Linear units measure the length of objects.

Finding the perimeter of regular shapes is easy. You simply measure one side of the shape and multiply that length by the number of sides in the shape. So to find the perimeter of this square, you would multiply 4 sides × 4 meters, which gives you 16 meters.

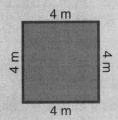

To find the perimeter of the pentagon below, you would multiply 5 sides × 3 inches, which gives you 15 inches.

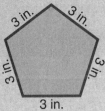

Irregular shapes do not have sides that are all the same length. Some sides may be short and some may be long. Here are some examples of irregular shapes.

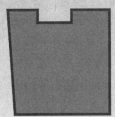

#50754—Leveled Texts for Mathematics: Measurement

Steps for Measuring the Perimeter of Irregular Shapes

Finding the perimeter of irregular shapes takes a few more steps than finding the perimeter of regular shapes. The perimeter will equal the sum of the segments around the outside of the shape.

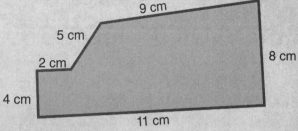

To find the perimeter of the irregular shape at right, follow these three steps.

1. Find the endpoints of each line segment on the shape.

2. Measure the distance of each segment.

3. Add the measurements together. This will give you the perimeter of the irregular shape.

On the shape above, the line segments measured 4 cm, 2 cm, 5 cm, 9 cm, 8 cm, and 11 cm. If you add those numbers together, you get a perimeter of 39 cm.

Measuring the Perimeter of Irregular Shapes in Our Daily Lives

Not all shapes in our world are regular shapes, so there are times when the perimeter of irregular shapes must be measured. Swimming pools are often irregular in shape. Therefore, the perimeter must be measured before the ground is broken and concrete is poured.

When building a home, the floor plans are most often irregular in shape. Therefore, contractors must be able to measure the perimeter of the home, as well.

You Try It

Find the perimeter of the following irregular shapes.

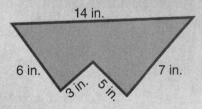

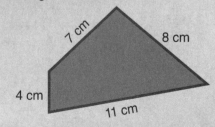

Measuring the Perimeter of Irregular Shapes

Look at the shape below. How could you find the perimeter of this shape?

Basic Facts

Perimeter is the distance around the outside of a figure. Because we are measuring length, we measure perimeter in linear units such as inches, centimeters, meters, feet, miles, or kilometers.

Finding the perimeter of regular shapes is easy because you simply measure one side of the shape and multiply that length by the number of sides in the shape. So to find the perimeter of this square, you would multiply 4 sides × 4 meters, which gives you 16 meters.

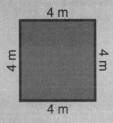

To find the perimeter of the pentagon below, you would multiply 5 sides × 3 inches, which gives you 15 inches.

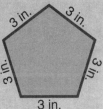

Irregular shapes, however, do not have sides that are all the same length; some sides may be short and some may be long. Here are some examples of irregular shapes.

 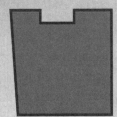 

107

Steps for Measuring the Perimeter of Irregular Shapes

Finding the perimeter of irregular shapes takes a few more steps than finding the perimeter of regular shapes. The perimeter will equal the sum of the segments around the outside of the figure.

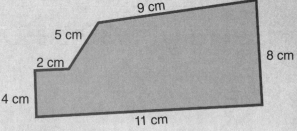

To find the perimeter of the irregular shape at right, follow these three steps:

1. Find the endpoints of each line segment on the shape.

2. Measure the distance of each segment.

3. Add all of the measurements together. This will give you the perimeter.

On the shape above, the line segments measure 4 cm, 2 cm, 5 cm, 9 cm, 8 cm, and 11 cm. If you add those measurements together, you get a perimeter of 39 cm.

Measuring the Perimeter of Irregular Shapes in Our Daily Lives

Not all shapes in our world are regular shapes, so there are instances when the perimeter of irregular shapes must be measured. Swimming pools are frequently irregular in shape and, therefore, it is imperative that the perimeter be accurately measured before the ground is broken and concrete is poured.

When building a home, the floor plans are most often irregular in shape, which means that contractors must be able to measure the perimeter of the home, as well.

You Try It

Find the perimeter of the following irregular shapes:

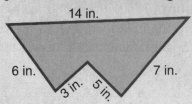

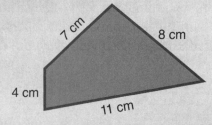

Measuring the Area of Irregular Shapes

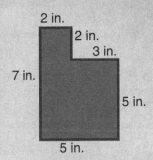

The area of this irregular shape is 29 inches². How can you tell? How can you prove this?

Basic Facts

What is area? It is the number of units that cover the inside of a shape. It is told in square units. Area only measures two-dimensional things, or flat shapes. It is easy to find the area of regular shapes. They have straight sides. All the sides and angles are the same. Equations can help. They let you find the area of regular shapes. They keep you from having to count the units inside. But, finding the area of irregular shapes takes more work. Irregular shapes have sides that are not the same lengths. They have angles that are not the same.

Steps for Measuring the Area of Irregular Shapes

1. Split the irregular shape into regular shapes. Make as many as you need.

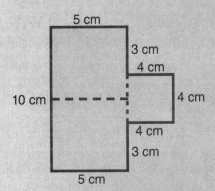

2. Look for any missing lengths on the new shapes. Use the lengths that are given. This can help you to find the missing ones.

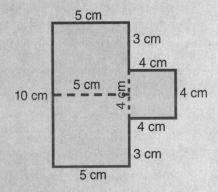

3. Find the area of the regular shapes you made.

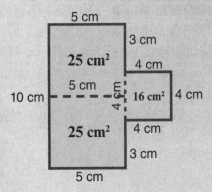

$$4 \text{ cm} \times 4 \text{ cm} = 16 \text{ cm}^2$$
$$2(5 \text{ cm} \times 5 \text{ cm}) = 50 \text{ cm}^2$$

4. Add up the areas of the new shapes. The total is the area of the irregular shape. Make sure you add them all.

$$16 \text{ cm}^2 + 50 \text{ cm}^2 = 66 \text{ cm}^2$$

109

Tips for Finding the Area of Irregular Shapes

Here are some tips. They can help find the area of irregular shapes:

- Split the irregular shape into more than one regular shape.

- Look carefully at every side. This will help you see any new lengths.

- Opposite sides are usually equal. This can be useful! You can find new lengths by looking at the opposite sides.

Measuring Area in Our Daily Lives

You may need to find the area of irregular shapes. Think about seeding your yard. You will need to know the area so you can get enough grass seed. Not all yards are regular shapes. Think of when you put mulch around plants. You need to know the area of the irregular shapes to buy enough mulch. Irregular shapes are all around us.

You Try It

Find the area of the shape below.

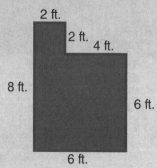

Measuring the Area of Irregular Shapes

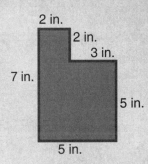

2 in.
2 in.
3 in.
7 in.
5 in.
5 in.

The area of this irregular shape is 29 inches². How can you prove that this is true?

Basic Facts

Area is the number of square units needed to cover the space inside a shape. Area only measures two-dimensional surfaces. Finding the area of regular shapes is easy. Regular shapes have straight sides. All of the sides are equal. And all the angles are equal. Equations can help you find the area of regular shapes. But finding the area of irregular shapes takes more work. Irregular shapes have sides that are different lengths. They have angles that are not equal.

Steps for Measuring the Area of Irregular Shapes

1. Divide the irregular figure into regular shapes.

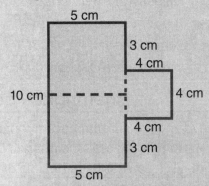

2. Look for any missing lengths on the new figure. Use the lengths that are given on the shape. This can help you to figure out what the missing ones are.

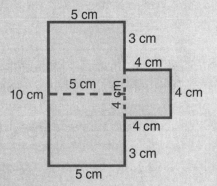

3. Find the area of the regular shapes you made.

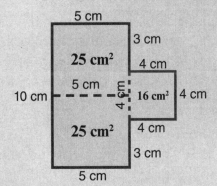

4. Add the areas of the regular shapes together. This will give you the total area of the irregular shape.

$$16 \text{ cm}^2 + 50 \text{ cm}^2 = 66 \text{ cm}^2$$

$$4 \text{ cm} \times 4 \text{ cm} = 16 \text{ cm}^2$$
$$2(5 \text{ cm} \times 5 \text{ cm}) = 50 \text{ cm}^2$$

(111)

Tips for Finding the Area of Irregular Shapes

Here are some tips. They can help find the area of irregular shapes.

- Divide the irregular figure into more than one regular shape.

- Look carefully at every side of the shape. This will help you see any new lengths.

- Opposite sides are usually equal. This can be useful! You can find new lengths by looking at the opposite sides of the shape.

Measuring Area in Our Daily Lives

There are times when you may need to find the area of irregular shapes. Think about seeding your yard. You will need to know the area so you can get enough grass seed. Not all yards are regular shapes. Think of when you put mulch around trees and plants. You need to know the area of the irregular shapes to buy enough mulch. Irregular shapes are all around us.

You Try It

Find the area of the shape below.

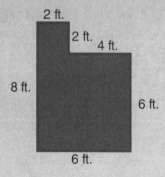

2 ft.

2 ft. 4 ft.

8 ft.

6 ft.

6 ft.

#50754—*Leveled Texts for Mathematics: Measurement*   © *Shell Education*

Measuring the Area of Irregular Shapes

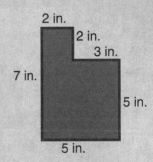

2 in.
2 in.
3 in.
7 in.
5 in.
5 in.

The area of this irregular shape is 29 inches². How can you prove that this is true?

Basic Facts

Area is the number of square units needed to cover the space inside a figure. Area only measures two-dimensional surfaces. Finding the area of regular shapes is easy. This is because regular shapes have straight sides. And all the sides and angles are equal. There are equations you can use to help you find the area of regular shapes. But finding the area of irregular shapes takes a bit more work. Irregular shapes have sides and angles that are not all equal.

Steps for Measuring the Area of Irregular Shapes

1. Divide the irregular shape into regular shapes.

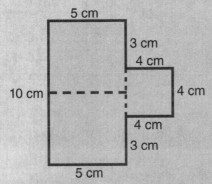

2. Look for any missing measurements on the new shape. Use the measurements provided on the shape to figure out what the missing measurements are.

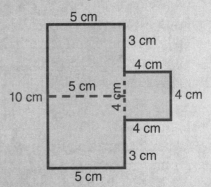

3. Find the area of the regular shapes you created.

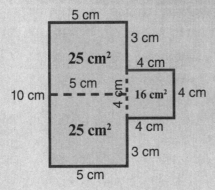

4. Add the areas of the regular shapes together. This will give you the total area of the irregular shape.

$$16 \text{ cm}^2 + 50 \text{ cm}^2 = 66 \text{ cm}^2$$

$$4 \text{ cm} \times 4 \text{ cm} = 16 \text{ cm}^2$$
$$2(5 \text{ cm} \times 5 \text{ cm}) = 50 \text{ cm}^2$$

Tips for Finding the Area of Irregular Shapes

Here are some tips for finding the area of irregular shapes.

- Divide the irregular shape into more than one regular shape to help you find the area.

- Look carefully at every side of the shape. This will help you find the new measurements once the shape is divided.

- Opposite sides are usually equal. So once you have divided the irregular shape into regular shapes, you can find the new measurements by looking at the opposite sides of the shape.

Measuring Area in Our Daily Lives

There are times when you may need to find the area of irregular shapes. If you are seeding your yard, you will need to know the area so you can get enough grass seed. And not all yards are regular shapes. Also when you place mulch around trees and other landscaping, you would need to know the area of the irregular shapes to buy enough mulch. Irregular shapes are all around us.

You Try It

Find the area of the shape below.

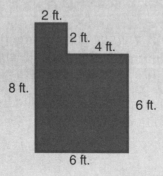

2 ft.
2 ft.
4 ft.
8 ft.
6 ft.
6 ft.

#50754—*Leveled Texts for Mathematics: Measurement*

Measuring the Area of Irregular Shapes

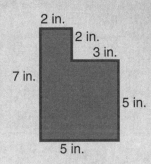

The area of this irregular shape is 29 inches². How can you prove that this is true?

Basic Facts

Area is the number of square units needed to cover the space inside a two-dimensional figure. Area only measures two-dimensional surfaces. Finding the area of regular shapes is easy because regular shapes have straight sides, of the side lengths and angle measures are equal. There are simple equations you can use to help you find the area of regular shapes, but finding the area of irregular shapes takes a bit more work. Irregular shapes have sides and angles that are not all equal.

Steps for Measuring the Area of Irregular Shapes

1. Divide the irregular shape into regular shapes.

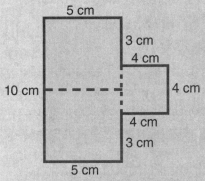

2. Look for any missing measurements on the new shape. Use the measurements provided on the shape to figure out what the missing measurements are.

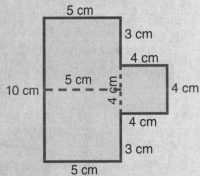

3. Find the area of each of the regular shapes you created.

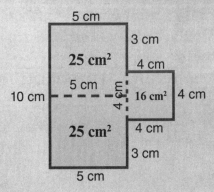

$$4 \text{ cm} \times 4 \text{ cm} = 16 \text{ cm}^2$$
$$2(5 \text{ cm} \times 5 \text{ cm}) = 50 \text{ cm}^2$$

4. Add the areas of the regular figures together to determine the total area of the irregular shape.

$$16 \text{ cm}^2 + 50 \text{ cm}^2 = 66 \text{ cm}^2$$

115

Tips for Finding the Area of Irregular Shapes

Here are some tips for finding the area of irregular shapes.

- Divide the irregular shape into more than one regular shape to help you find the area.

- Look carefully at every side of the shape to help you find the new measurements once the shape is divided.

- Opposite sides are usually equal, so once you have divided the irregular shape into regular shapes, you can often find the new measurements by looking at the opposite sides of the shape.

Measuring Area in Our Daily Lives

There are times when you may need to find the area of irregular shapes. Not all yards are regular shapes, but if you are seeding your yard, you will need to know the area so you can get enough grass seed. Also, when you place mulch around trees and other landscaping, you need to know the area of the irregular shapes to buy enough mulch. Irregular shapes are all around us.

You Try It

Find the area of the irregular shape below. Remember to consider all of the tips for finding the area of irregular shapes.

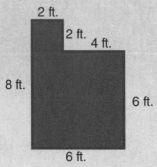

Measuring the Volume of Solids and Liquids

You have a basketball. And you have a tennis ball. Which takes up more space? How do you know?

Basic Facts

How much space a thing takes is called **volume**. Volume is shown in cubic units. This is because things take up three dimensions. They have length. They have width. And they have height.

A car's gas tank may hold 20 gallons of gas. An ice cube may have a volume of 2 cubic centimeters. Gas is a liquid. Ice is a solid. Both have volume. But you find it in different ways.

Measuring the Volume of a Liquid

How do you show the volume of a liquid? You can use liters. This is in the metric system. What does the English units system use? It uses quarts. It uses pints. It uses gallons.

We measure liquids with tools. We use graduated cylinders. We can use beakers. Or we use liquid measuring cups. These can show how much space a liquid takes. You fill the tool. Then you read the lines. They show how much is filled.

Beakers often show milliliters. So do graduated cylinders. We write the unit as mL. Liquid measuring cups can use fluid ounces. Or they may use cups.

Measuring the Volume of a Regular Solid

Finding the volume of regular-shaped things is easy. You use a formula. Volume is length × width × height. Think of your math book. How can you find its volume? You use a ruler. You measure the length. You find width. Then the height. Then you multiply them.

5 inches

11 inches

} 2 inches

The book above has a length of 11 inches. It has a width of 5 inches. And it has a height of 2 inches. You multiply 11 × 5 × 2. You get 110. This means that the book takes up 110 inches cubed of space. That can be written 110 in.[3]

117

Measuring the Volume of an Irregular Solid

Irregular solids are not easy to measure. They are things like rocks. Think of a rock. It has places that are rough. It has places that push out. There is no one length. There is no one height. Or the length, width, and height cannot be measured easily. The shape is irregular. We can use water displacement. This lets us measure their volume. You put the thing in water. Then you check the water level. You see how much it changes. Here are simple steps. Follow them to measure the volume of an irregular solid.

1. Fill a beaker half way. The water should be able to cover the solid.

2. Measure the water. Record the water level in the beaker.

3. Put the item into the water.

4. Measure the water. Record the new water level.

5. Subtract the original water level from the new water level. This will give you the volume of the irregular solid.

Measuring Volume in Our Daily Lives

Think about using a recipe. You must know how to measure volume. This is very important in a baking recipe. A cookie recipe may call for $\frac{1}{2}$ cup of vegetable oil. It may need 3 tablespoons of water. What if you add the wrong volume of liquids? The cookies may not come out right. They might be very dry. Or they might be too wet.

You Try It

What method would you use to find the volume of the items below?

- a sea shell
- a brick
- a glass of milk

Measuring the Volume of Solids and Liquids

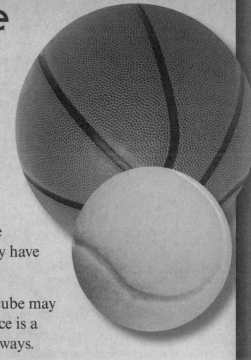

Which takes up more space? Does a basketball or a tennis ball? How do you know?

Basic Facts

The amount of space that a thing takes is called **volume**. Volume is shown in cubic units. This is because things take up three dimensions. They have length. They have width. And they have height.

A car's gas tank may hold 20 gallons of gas. An ice cube may have a volume of 2 cubic centimeters. Gas is a liquid. Ice is a solid. Each has volume. But you find them in different ways.

Measuring the Volume of a Liquid

The volume of a liquid is measured in liters. This is in the metric system. What does the English units system use? It uses fluid ounces. It uses quarts. It uses pints. And it uses gallons.

For liquids, we use tools to measure. We use things like graduated cylinders and beakers. Or we use liquid measuring cups.

Beakers and graduated cylinders are often measured in milliliters. We write the unit as mL. Liquid measuring cups often use fluid ounces or cups.

Measuring the Volume of a Regular Solid

Finding the volume of regular-shaped items is easy to do. You use the formula length × width × height. Think of your math book. How would you measure its volume? You would measure the length, width, and height of the book using a ruler.

5 inches
11 inches
2 inches

The book pictured here has a length of 11 inches. It has a width of 5 inches. And it has a height of 2 inches. If you multiply 11 × 5 × 2, you get 110 inches cubed. This means that the book takes up 110 inches cubed of space or 110 in.[3]

(119)

Measuring the Volume of an Irregular Solid

Water displacement is used to measure the volume of an irregular solid. Irregular solids are items such as rocks. Their length, width, and height cannot be measured accurately. There are simple steps you can follow to measure the volume of an irregular solid.

1. Fill a beaker half way. The water should be able to cover the solid.

2. Measure and record the water level in the beaker.

3. Put the irregular solid in the beaker.

4. Measure and record the new water level.

5. Subtract the original water level from the new water level. This will give you the volume of the irregular solid.

Measuring Volume in Our Daily Lives

Think about using a recipe. You must know how to measure volume. This is very important in a baking recipe. A cookie recipe may call for $\frac{1}{2}$ cup of vegetable oil. It may need 3 tablespoons of water. What if you add the wrong volume of liquids? The cookies may not come out right. They might be very dry. Or they might be too wet.

You Try It

What method would you use to find the volume of the items below?

- a sea shell
- a brick
- a glass of milk

Measuring the Volume of Solids and Liquids

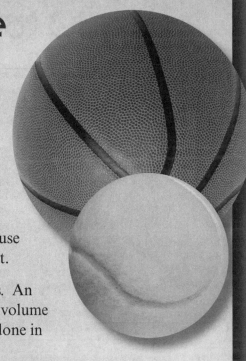

Which takes up more space: a basketball or a tennis ball? How do you know?

Basic Facts

The amount of space that an object occupies is called **volume**. Volume is measured in cubic units. This is because you measure three dimensions: length, width, and height.

A car's gas tank may be able to hold 20 gallons of gas. An ice cube may have a volume of 2 cubic centimeters. The volume of both solids and liquids can be measured. But this is done in different ways.

Measuring the Volume of a Liquid

The volume of a liquid is measured in liters in the metric system. In the English units system, it is measured in fluid ounces, quarts, pints, and gallons.

To measure the volume of a liquid, we use tools such as graduated cylinders, beakers, or liquid measuring cups.

Most often, beakers and graduated cylinders are measured in milliliters. We write the unit as mL. Liquid measuring cups often measure in fluid ounces or cups.

Measuring the Volume of a Regular Solid

Measuring the volume of regular-shaped items, such as a book, is simple to do. You use the formula length × width × height. To measure the volume of your math book, you would measure the length, width, and height of the book using a ruler.

5 inches 11 inches } 2 inches

The book pictured above has a length of 11 inches, a width of 5 inches, and a height of 2 inches. So if you multiply 11 × 5 × 2, you would get 110 inches cubed. This means that the book takes up 110 inches cubed or 110 in.³ of space.

 #50754—Leveled Texts for Mathematics: Measurement

Measuring the Volume of an Irregular Solid

Water displacement is used to measure the volume of an irregular solid. Irregular solids are items such as rocks, whose length, width, and height cannot be measured accurately. There are simple steps you can follow to measure the volume of an irregular solid.

1. Fill a beaker half way, so that it would cover the solid.

2. Measure and record the water level in the beaker.

3. Place the object in the beaker.

4. Measure and record the new water level.

5. Subtract the original water level from the new water level. This will give you the object's volume.

Measuring Volume in Our Daily Lives

Think about following a recipe. It is very important to understand how to measure volume. This is especially important in a baking recipe. For example, a cookie recipe may call for $\frac{1}{2}$ cup of vegetable oil and 3 tablespoons of water. If you do not add the correct volume of liquids, the cookies may not come out right. They might be very dry. Or they might be too runny.

You Try It

What method would you use to find the volume of the items below?

- a sea shell
- a brick
- a glass of milk

Measuring the Volume of Solids and Liquids

Which takes up more space: a basketball or a tennis ball? How did you make that determination?

Basic Facts

The amount of space that an object occupies is called **volume**. Volume is measured in cubic units because you measure three dimensions: length, width, and height.

A car's gas tank may be able to hold 20 gallons of gas, while an ice cube may have a volume of 2 cubic centimeters. The volume of both solids and liquids can be measured, but they are measured in different ways.

Measuring the Volume of a Liquid

The volume of a liquid is measured in liters in the metric system, while in the English units system it is measured in fluid ounces, quarts, pints, and gallons.

To measure the volume of a liquid, we use tools such as graduated cylinders, beakers, or liquid measuring cups.

Most often, beakers and graduated cylinders are measured in milliliters, which can be written as mL. Liquid measuring cups often measure in fluid ounces or cups.

Measuring the Volume of a Regular Solid

Measuring the volume of regular-shaped items, such as a book, is simple to do. You use the formula length × width × height. Thus, to measure the volume of your math book, you would measure the length, width, and height of the book using a ruler and then do the multiplication problem.

5 inches
11 inches
2 inches

The book pictured above has a length of 11 inches, a width of 5 inches, and a height of 2 inches. So, if you multiply 11 × 5 × 2, you would get 110 inches cubed. This means that the book takes up 110 inches cubed of space, or 110 in.³

123

Measuring the Volume of an Irregular Solid

Water displacement is used to measure the volume of an irregular solid. Irregular solids are items such as rocks, whose length, width, and height cannot be measured accurately. There are simple steps you can follow to measure the volume of an irregular solid.

1. Fill a beaker half way, so that it would cover the solid.

2. Measure and record the water level in the beaker.

3. Place the object in the beaker.

4. Measure and record the new water level.

5. Subtract the original water level from the new water level. This will give you the solid's volume.

Measuring Volume in Our Daily Lives

Think about following a recipe. It is very important to understand how to measure volume, especially in a baking recipe. For example, a cookie recipe may call for $\frac{1}{2}$ cup of vegetable oil and 3 tablespoons of water, but if you do not add the correct volume of liquids, the cookies may not come out right. They might be very dry or they might be too runny.

You Try It

What method would you use to find the volume of the items below?

- a sea shell
- a brick
- a glass of milk

Converting Volume

You have a fish tank. It has directions. You read them. They say, "Fill with water. Do Not Use More Than 1.5 Gallons!" The fish need clean water. But you only have one cup. It is in fluid ounces. You want to make sure that the water is right. You do not want too much. And you do not want too little. You must convert. You must convert from gallons to ounces.

Knowing how to convert volume would help. There are different rules. They will help. They let you convert from one unit to another.

Converting Volume in the English Standard Unit of Measurement

Measurement	Conversion
1 pint	16 fluid ounces
1 quart	2 pints
1 gallon	4 quarts or 8 pints

English Units Conversion Rule One: To convert a small unit to a bigger one, divide.

You have 16 pints of milk. How many gallons is that? There are 8 pints in 1 gallon. So you divide 16 by 8. This gives you 2. 16 pints is 2 gallons.

$$16 \text{ pints} \div \frac{1 \text{ gallon}}{8 \text{ pints}} = 2 \text{ gallons}$$

English Units Conversion Rule Two: To convert a big unit to a smaller one, multiply.

You have 5 gallons. How many quarts is that? There are 4 quarts in 1 gallon. You multiply 5 by 4. This gives you 20. 5 gallons is 20 quarts.

$$5 \text{ gallons} \times \frac{4 \text{ quarts}}{1 \text{ gallon}} = 20 \text{ quarts}$$

Converting Volume in the Metric System

Converting units in metric can be easy. Metric is based on groups of 10. You multiply or divide by 10s. Or you use multiples of 10. That makes the math easy.

What is a base unit? It is what you are using to measure. The base unit of volume is liter. This is in metric. Big things use kiloliters. For small things, use milliliters.

Converting Volume in the Metric System (cont.)

Metric Unit	Unit Multiple
kiloliters	1,000 (one thousand)
hectoliters	100 (one hundred)
decaliters	10 (ten)
Base Unit: liter	1
deciliters	0.1 (one-tenth)
centiliters	0.01 (one-hundredth)
milliliters	0.001 (one-thousandth)

Metric Conversion Rule One: To convert a small unit to a bigger one, divide.

Tina has 6,500 milliliters of soda. Milliliters are small. She wants to see how many liters that is. Liters are bigger. She divides by 1,000. Why? She looked at the chart. Milliliters are three units apart from liters. She has 6.5 liters.

$$6{,}500 \text{ milliliters} \div \frac{1 \text{ liter}}{1{,}000 \text{ milliliters}} = 6.5 \text{ liters}$$

Metric Conversion Rule Two: To convert from a big unit to a smaller one, multiply.

Ray has some milk. He has 9.7 hectoliters. How many milliliters is that? Milliliters are very small. He can convert. He uses the chart. He multiplies by 100,000. There are 970,000 milliliters in 9.7 hectoliters. That is a lot of milliliters!

$$9.7 \text{ hectoliters} \times \frac{100{,}000 \text{ milliliters}}{1 \text{ hectoliter}} = 970{,}000 \text{ milliliters}$$

Converting Volume in Our Daily Lives

Pharmacists must convert volume. They have to keep people safe. They may have a large bottle of medicine. It holds a quart. But they only need to fill a small bottle. They need to fill 45 milliliters. They do not want to give too much. That could make someone sick. And it would cost more. They do not want to give too little. That would not help. They must convert volume. This will let them give the patient the right amount.

You Try It

Convert these units:

20 pints = _____ gallons

93 liters = _____ milliliters

Converting Volume

You have a fish tank. It has directions. They read, "2 Gallon Tank—Do Not Go Over 1.5 Gallons!" It is your job to clean it. But you only have one cup. It is in fluid ounces. You want to make sure that the water is right. You do not want too much water or too little water. You must convert gallons to ounces.

Knowing how to convert volume would help you. There are different rules. They will help to convert from one unit to another.

Converting Volume in the English Standard Unit of Measurement

Measurement	Conversion
1 pint	16 fluid ounces
1 quart	2 pints
1 gallon	4 quarts or 8 pints

English Units Conversion Rule One: To convert a smaller unit to a bigger one, divide.

You have 16 pints of milk. You want to see how many gallons that is. You divide 16 by 8. This is because there are 8 pints in 1 gallon. This gives you 2. 16 pints is 2 gallons.

$$16 \text{ pints} \div \frac{1 \text{ gallon}}{8 \text{ pints}} = 2 \text{ gallons}$$

English Units Conversion Rule Two: To convert a larger unit to a smaller one, multiply.

You have 5 gallons. You want to see how many quarts that is. You multiply 5 by 4. This is because there are 4 quarts in 1 gallon. This gives you 20. 5 gallons is 20 quarts.

$$5 \text{ gallons} \times \frac{4 \text{ quarts}}{1 \text{ gallon}} = 20 \text{ quarts}$$

Converting Volume in the Metric System

Converting units in metric can be easy. Metric is based on groups of 10. You have to multiply by 10s. Or you need to divide by 10s. This will make units either bigger or smaller.

The base unit is the unit you are using to measure. The base unit of volume in the metric system is liter. Big things are measured in kiloliters. The smallest volumes use milliliters.

Converting Volume in the Metric System (cont.)

Metric Unit	Unit Multiple
kiloliters	1,000 (one thousand)
hectoliters	100 (one hundred)
decaliters	10 (ten)
Base Unit: liter	1
deciliters	0.1 (one-tenth)
centiliters	0.01 (one-hundredth)
milliliters	0.001 (one-thousandth)

Metric Conversion Rule One: To convert a smaller unit to a bigger one, divide.

Tina has 6,500 milliliters of soda. She wants to see how many liters that is. She divides by 1,000. This is because milliliters are three units apart from liters. This is 6.5 liters.

$$6{,}500 \text{ milliliters} \div \frac{1 \text{ liter}}{1{,}000 \text{ milliliters}} = 6.5 \text{ liters}$$

Metric Conversion Rule Two: To convert from a larger unit to a smaller unit, multiply.

Ray has 9.7 hectoliters of milk. He can convert that to milliliters. He can multiply by 100,000. This would convert the 9.7 hectoliters to 970,000 milliliters.

$$9.7 \text{ hectoliters} \times \frac{100{,}000 \text{ milliliters}}{1 \text{ hectoliter}} = 970{,}000 \text{ milliliters}$$

Converting Volume in Our Daily Lives

Pharmacists must convert volume. They do this to fill prescriptions. They may have a large bottle of medicine. It holds a quart. But they may only need to fill a small bottle. The bottle might hold 45 milliliters. They need to know how to convert volume. This will let them give the patient the right amount.

You Try It

Convert these units of measurement.

20 pints = _____ gallons

93 liters = _____ milliliters

Converting Volume

The directions with your fish bowl read "2 Gallon Tank—Do Not Exceed 1.5 Gallons!" It is your job to clean the fish tank. And the only cup you have to refill the water is in fluid ounces. In order to make sure that the fish tank is at the right level, you must convert from gallons to ounces.

Knowing how to convert volume would help you solve this problem. There are different rules you can follow to convert from one unit of volume to another.

Converting Volume in the English Standard Unit of Measurement

Measurement	Conversion
1 pint	16 fluid ounces
1 quart	2 pints
1 gallon	4 quarts or 8 pints

English Units Conversion Rule One: To convert a smaller unit to a larger unit, you should divide.

You have 16 pints of milk. How many gallons is that? You divide 16 pints by 8 because there are 8 pints in 1 gallon. This gives you 2 gallons.

$$16 \text{ pints} \div \frac{1 \text{ gallon}}{8 \text{ pints}} = 2 \text{ gallons}$$

English Units Conversion Rule Two: To convert a larger unit to a smaller unit, you should multiply.

You have 5 gallons, and you want to see how many quarts that would equal. You multiply 5 gallons by 4 because there are 4 quarts in 1 gallon. This gives you 20 quarts.

$$5 \text{ gallons} \times \frac{4 \text{ quarts}}{1 \text{ gallon}} = 20 \text{ quarts}$$

Converting Volume in the Metric System

Converting units in the metric system can be easy. This is because the metric system is based on multiples of 10. You have to multiply or divide by multiples of 10 to make units either larger or smaller.

The base unit is the unit you are using to measure. So the base unit of volume in the metric system is liter. This means that large volumes are measured in kiloliters and the smallest volumes in milliliters.

Converting Volume in the Metric System *(cont.)*

Metric Unit	Unit Multiple
kiloliters	1,000 (one thousand)
hectoliters	100 (one hundred)
decaliters	10 (ten)
Base Unit: liter	1
deciliters	0.1 (one-tenth)
centiliters	0.01 (one-hundredth)
milliliters	0.001 (one-thousandth)

Metric Conversion Rule One: To convert a smaller unit to a larger unit, you must divide.

Tina has 6,500 milliliters of soda. She wants to see how many liters that is. She divides by 1,000. This is because milliliters are three units apart from liters. This results in 6.5 liters.

$$6{,}500 \text{ milliliters} \div \frac{1 \text{ liter}}{1{,}000 \text{ milliliters}} = 6.5 \text{ liters}$$

Metric Conversion Rule Two: To convert from a larger unit to a smaller unit, you must multiply.

Ray has 9.7 hectoliters of milk. He could convert that to milliliters by multiplying by 100,000. By doing so, he would convert the 9.7 hectoliters to 970,000 milliliters.

$$9.7 \text{ hectoliters} \times \frac{100{,}000 \text{ milliliters}}{1 \text{ hectoliter}} = 970{,}000 \text{ milliliters}$$

Converting Volume in Our Daily Lives

Pharmacists must convert volume. They do this to fill prescriptions. They may have a large bottle that holds a quart of medicine. But they only need to give someone a bottle that measures 45 milliliters. They need to know how to convert volume to give the patient the right amount of medicine.

You Try It

Convert the units of measurement below.

20 pints = _____ gallons

93 liters = _____ milliliters

Converting Volume

The directions with your fish bowl read, "2 Gallon Tank—Do Not Exceed 1.5 Gallons!" It is your job to clean the fish tank and the only cup you have to refill the water is in fluid ounces. In order to make sure that the fish tank is at the right level, you must convert from gallons to ounces.

Knowing how to convert volume would help you solve this problem. There are different rules you can follow to convert from one unit of volume to another.

Converting Volume in the English Standard Unit of Measurement

Measurement	Conversion
1 pint	16 fluid ounces
1 quart	2 pints
1 gallon	4 quarts or 8 pints

English Units Conversion Rule One: To convert a smaller unit to a larger unit, you should divide.

You have 16 pints of milk and you want to see how many gallons that is. You divide 16 pints by 8 because there are 8 pints in 1 gallon. This gives you 2 gallons.

$$16 \text{ pints} \div \frac{1 \text{ gallon}}{8 \text{ pints}} = 2 \text{ gallons}$$

English Units Conversion Rule Two: To convert a larger unit to a smaller unit, you should multiply.

You have 5 gallons, and you want to see how many quarts that would equal. You multiply 5 gallons by 4 because there are 4 quarts in 1 gallon. This gives you 20 quarts.

$$5 \text{ gallons} \times \frac{4 \text{ quarts}}{1 \text{ gallon}} = 20 \text{ quarts}$$

Converting Volume in the Metric System

Converting units in the metric system can be easy since the metric system is based on multiples of 10. You have to multiply or divide by multiples of 10 to make units either larger or smaller.

The base unit is the unit you are using to measure. Since the base unit of volume in the metric system is liter, large volumes are measured in kiloliters and the smallest volumes in milliliters.

Converting Volume in the Metric System (cont.)

Metric Unit	Unit Multiple
kiloliters	1,000 (one thousand)
hectoliters	100 (one hundred)
decaliters	10 (ten)
Base Unit: liter	1
deciliters	0.1 (one-tenth)
centiliters	0.01 (one-hundredth)
milliliters	0.001 (one-thousandth)

Metric Conversion Rule One: To convert a smaller unit to a larger unit, you must divide.

Tina has 6,500 milliliters of soda, but she wants to see how many liters that is. Because milliliters are three units apart from liters, she divides by 1,000. This results in 6.5 liters.

$$6{,}500 \text{ milliliters} \div \frac{1 \text{ liter}}{1{,}000 \text{ milliliters}} = 6.5 \text{ liters}$$

Metric Conversion Rule Two: To convert from a larger unit to a smaller unit, you must multiply.

Ray has 9.7 hectoliters of milk. He could convert that to milliliters by multiplying by 100,000. By doing so, he would convert the 9.7 hectoliters to 970,000 milliliters.

$$9.7 \text{ hectoliters} \times \frac{100{,}000 \text{ milliliters}}{1 \text{ hectoliter}} = 970{,}000 \text{ milliliters}$$

Converting Volume in Our Daily Lives

Pharmacists must convert volume measurements to fill prescriptions. They may have a large bottle that holds a quart of medicine, but they only need to dispense enough for a bottle that measures 45 milliliters. They need to know how to convert volume to give the patient the right amount of medicine.

You Try It

Convert the units of measurement below.

20 pints = _____ gallons

3 liters = _____ milliliters

Measuring Surface Area

What is the surface area of Earth? It is 510 million square kilometers!

Basic Facts

Surface area is different than area. **Surface area** tells us how much space a thing takes up on the outside. It is the total area of all the outer sides.

Surface area is shown in square units. How do you find it? You add up the areas of the faces. First you find the area of each face. Then you add them.

There are equations you can use. They show the surface area. They make it easy. Each shape has one.

Shape	Equation	Picture of Shape
cube	$6s^2$ 1. Square the sides. 2. Multiply the number by 6.	4 m, 4 m, 4 m
sphere	$4\pi r^2$ 1. Square the radius. 2. Multiply the number by 3.14, or π. 3. Multiply your answer by 4.	7 in.
rectangular prism	$2(lw + lh + wh)$ 1. Multiply length × width. 2. Multiply length × height. 3. Multiply width × height. 4. Add the products together. 5. Multiply by 2.	5 cm, 4 cm, 8 cm

#50754—*Leveled Texts for Mathematics: Measurement*

Working Out the Surface Area

Here are some examples. They show how to use the equations.

Look at the cube at right. It has 6 sides. Each is 4 meters long. Look at the chart. Find *cube*. Use the equation $6s^2$. Follow the steps. First, square 4 meters. That is 16 m². Then multiply that by 6. This is for the six sides. The answer is 96 m².

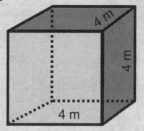

Look at the rectangular prism at right. It is 5 cm long. It is 4 cm high. It is 8 cm wide. Use the equation $2(lw + lh + wh)$. First, multiply the length by the width. This gives 40 cm². Then, multiply the length by the height. This gives 20 cm². Then, multiply the width by the height. This is 32 cm². Add the three together. You get 92 cm². Multiply that by 2. This is 184 cm². This is the answer. The surface area is 184 cm².

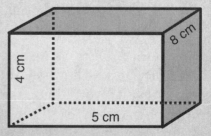

Measuring Surface Area in Our Daily Lives

Who uses surface area? Scientists do. They look at planets. They compare the surface areas. They look at Earth's surface area. They know which parts of Earth have water. They know if the amount of water is changing. They look at the amount over time.

You Try It

Find the surface area of the shapes below.

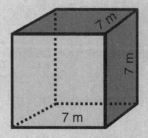

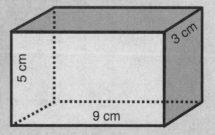

Measuring Surface Area

What is the surface area of Earth? It is 510 million square kilometers!

Basic Facts

Surface area and area are different. **Surface area** tells us how much space a thing takes up on the outside. It measures the total area of all the outer sides.

Surface area is told in square units. How do you find it? You add up the areas of the faces of a shape. First you find the area of each face. Then you add the areas together.

There are different equations you can use. These will help to find surface area for shapes.

Shape	Equation	Picture of Shape
cube	$6s^2$ **1.** Square the sides. **2.** Multiply the number by 6.	4 m, 4 m, 4 m
sphere	$4\pi r^2$ **1.** Square the radius. **2.** Multiply the number by 3.14, or π. **3.** Multiply your answer by 4.	7 in.
rectangular prism	$2(lw + lh + wh)$ **1.** Multiply length × width. **2.** Multiply length × height. **3.** Multiply width × height. **4.** Add the products together. **5.** Multiply by 2.	5 cm, 4 cm, 8 cm

#50754—Leveled Texts for Mathematics: Measurement

Working Out the Surface Area

Here are some examples. They show how to use the equations.

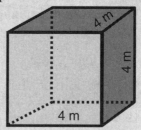

Look at the cube at right. It has 6 sides. Each one is 4 meters long. The equation for the xsurface area of a cube is $6s^2$. First square 4 meters. That is 16 meters squared. Then multiply that by 6. This is for the six sides. The answer is 96 m².

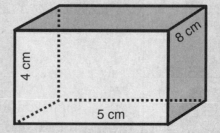

Look at the rectangular prism at right. It has a length of 5 cm. It has a height of 4 cm. It has a width of 8 cm. Use the equation $2(lw + lh + wh)$. First, multiply the length by the width. This gives you 40 cm². Then, multiply the length by the height. This gives 20 cm². Then, multiply the width by the height. This is 32 cm². Add the three together. You will then get a total of 92 cm². Multiply that by 2. This is the answer. The surface area is 184 cm².

Measuring Surface Area in Our Daily Lives

Scientists measure the surface area of different planets. They want to compare the surface areas of planets. They compare Earth's surface area to other planets. Astronomers can see which parts of Earth are covered in water and which parts are changing over time.

You Try It

Find the surface area of the shapes below.

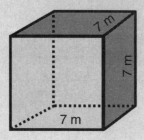

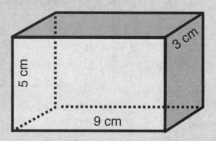

Measuring Surface Area

Did you know that the total surface area of Earth is 510 million square kilometers?

Basic Facts

Surface area and area are different. **Surface area** tells us how much space an object takes up on the outside. It measures the total area of the outer sides an object.

Surface area is measured in square units. When you find surface area, you are finding the sum of the areas of the faces on a shape. This means that surface area is found by first finding the area of each face on a three-dimensional shape, then adding the areas together.

There are different equations you can use to measure surface area for various shapes.

Shape	Equation	Picture of Shape
cube	$6s^2$ 1. Square the sides. 2. Multiply the number by 6.	4 m, 4 m, 4 m
sphere	$4\pi r^2$ 1. Square the radius. 2. Multiply the number by 3.14, or π. 3. Multiply your answer by 4.	7 in.
rectangular prism	$2(lw + lh + wh)$ 1. Multiply length × width. 2. Multiply length × height. 3. Multiply width × height. 4. Add the products together. 5. Multiply by 2.	5 cm, 4 cm, 8 cm

Working Out the Surface Area

Here are some examples of how to find the surface area of a shape.

The cube at right has sides that each measures 4 meters. The equation to find the surface area of a cube is $6s^2$. First square 4 meters, which equals 16 meters squared. Then, multiply that answer by 6. This is because there are six sides on a cube. The surface area of the cube at right is 96 m².

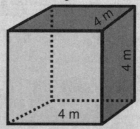

The rectangular prism at right has a length of 5 cm, a height of 4 cm, and a width of 8 cm. The equation to find the surface area of a rectangular prism is $2(lw + lh + wh)$. First, multiply the length, 5 cm, by the width, 8 cm. This gives you 40 cm². Then, multiply the length by the height to give you 20 cm². Finally, multiply the width by the height, which gives you 32 cm². Add the three products together. You will then get a total of 92 cm². Multiply that by 2 to get the surface area of 184 cm².

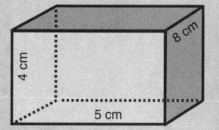

Measuring Surface Area in Our Daily Lives

Scientists measure the surface area of different planets in our solar system to compare them. Astronomers compare Earth's surface area to that of other planets. Knowing the surface area lets them know which parts of Earth are covered with water and which parts are changing over time.

You Try It

Find the surface area of the shapes below.

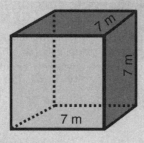

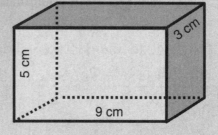

Measuring Surface Area

Did you know that the total surface area of Earth is 510 million square kilometers?

Basic Facts

Surface area and area are different. **Surface area** tells us how much space an object takes up on the outside, which means it measures the total area of all the outer sides of an object.

Surface area is measured in square units. When you find surface area, you are finding the sum of the areas of the faces of a three-dimensional shape, which means that surface area is found by first finding the area of each face on a three-dimensional shape, then adding the areas together.

There are different equations you can use to measure surface area for various three-dimensional shapes.

Shape	Equation	Picture of Shape
cube	$6s^2$ 1. Square the sides. 2. Multiply the number by 6.	*(cube with sides 4 m, 4 m, 4 m)*
sphere	$4\pi r^2$ 1. Square the radius. 2. Multiply the number by 3.14, or π. 3. Multiply your answer by 4.	*(sphere with 7 in. diameter)*
rectangular prism	$2(lw + lh + wh)$ 1. Multiply length × width. 2. Multiply length × height. 3. Multiply width × height. 4. Add the products together. 5. Multiply by 2.	*(rectangular prism with 4 cm, 8 cm, 5 cm)*

Working Out the Surface Area

Here are some examples of how to find the surface area of a shape.

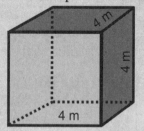

The cube at right has sides that each measures 4 meters. The equation to find the surface area of a cube is $6s^2$. First you square 4 meters, which equals 16 meters squared. Next, multiply that answer by 6 to account for the six sides on a cube. The surface area of the cube at right is 96 m².

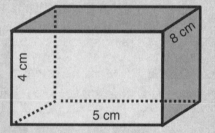

The rectangular prism at right has a length of 5 cm, a height of 4 cm, and a width of 8 cm. The equation to find the surface area of a rectangular prism is $2(lw + lh + wh)$. First, multiply the length, 5 cm, by the width, 8 cm, which totals 40 cm². Then, multiply the length by the height to give you 20 cm². Next, multiply the width by the height, which gives you 32 cm². Final, add the three products together to get a total of 92 cm² and multiply that by 2 to get a total surface area of 184 cm².

Measuring Surface Area in Our Daily Lives

Astronomers measure the surface area of various planets in our solar system in order to compare them to each other and learn about the characteristics of each planet. Astronomers also compare Earth's surface area over time in order to identify which parts of Earth are covered in water and which parts are altering.

You Try It

Find the surface area of the shapes below.

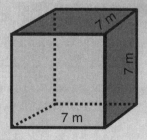

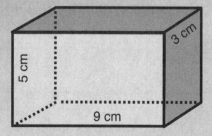

References Cited

August, D. and T. Shanahan (Eds). 2006. Developing literacy in second-language learners: Report of the National Literacy Panel on language-minority children and youth. Mahwah, NJ: Lawrence Erlbaum Associates, Inc.

Common Core State Standards Initiative. 2010. *The standards: Language arts.* (Accessed October 2010.) http://www.corestandards.org/the-standards/languagearts.

Marzano, R., D. Pickering, and J. Pollock. 2001. *Classroom instruction that works.* Alexandria, VA: Association for Supervision and Curriculum Development.

Tomlinson, C.A. 2000. *Leadership for Differentiating Schools and Classrooms.* Alexandria, VA: Association for Supervision and Curriculum Development.

Vygotsky, L.S. 1978. *Mind and society: The development of higher mental processes.* Cambridge, MA: Harvard University Press.

Contents of Teacher Resource CD

NCTM Mathematics Standards

The National Council of Teachers of Mathematics (NCTM) standards are listed in the chart on page 20, as well as on the Teacher Resource CD: *nctm.pdf*. TESOL standards are also included: *TESOL.pdf*

Text Files

The text files include the text for all four levels of each reading passage. For example, the Measuring the Length of Objects text (pages 21–28) is the *measuring_length.doc* file.

PDF Files

The full-color PDFs provided are each eight pages long and contain all four levels of a reading passage. For example, the Measuring the Length of Objects PDF (pages 21–28) is the *measuring_length.pdf* file.

Text Title	Text File	PDF
Measuring the Length of Objects	measuring_length.doc	measuring_length.pdf
Measuring the Weight of Objects	measuring_weight.doc	measuring_weight.pdf
Measuring Time	measuring_time.doc	measuring_time.pdf
Measuring Temperature	measuring_temperature.doc	measuring_temperature.pdf
Measuring Angles	measuring_angles.doc	measuring_angles.pdf
Estimating Measurements	estimating_measurements.doc	estimating_measurements.pdf
Converting Length	converting_length.doc	converting_length.pdf
Converting Weight	converting_weight.doc	converting_weight.pdf
Measuring the Perimeter of Regular Shapes	perimeter_regular.doc	perimeter_regular.pdf
Measuring the Area of Regular Shapes	area_regular.doc	area_regular.pdf
Measuring the Perimeter of Irregular Shapes	perimeter_irregular.doc	perimeter_irregular.pdf
Measuring the Area of Irregular Shapes	area_irregular.doc	area_irregular.pdf
Measuring the Volume of Solids and Liquids	measuring_volume.doc	measuring_volume.pdf
Converting Volume	converting_volume.doc	converting_volume.pdf
Measuring Surface Area	surface_area.doc	surface_area.pdf

JPEG Files

Key mathematical images found in the book are also provided on the Teacher Resource CD.

Word Documents of Texts

- Change leveling further for individual students.
- Separate text and images for students who need additional help decoding the text.
- Resize the text for visually impaired students.

Full-Color PDFs of Texts

- Create overhead transparencies or color copies to display on a document projector.
- Project texts on an interactive whiteboard or other screen for whole-class review.
- Read texts online.
- Email texts to parents or students at home.

JPEGs of Mathematical Images

- Display as visual support for use with whole class or small-group instruction.

142

Notes

#50754—*Leveled Texts for Mathematics: Measurement*

Notes